Servicing TV and Video Equipment

Servicing TV
and
Video Equipment

Eugene Trundle

MSERT MRTS MISTC

Newnes
An imprint of Butterworth-Heinemann Ltd
Linacre House, Jordan Hill, Oxford OX2 8DP

 PART OF REED INTERNATIONAL BOOKS

OXFORD   LONDON   BOSTON
MUNICH   NEW DELHI   SINGAPORE   SYDNEY
TOKYO   TORONTO   WELLINGTON

First published 1989
Reprinted 1990 (twice), 1991, 1992, 1993

British Library Cataloguing in Publication Data
Trundle, Eugene.
 Servicing TV and video equipment.
 1. Television equipment. Receivers. Maintenance &
 Repair. Manuals   2. Video equipment. Maintenance &
 Repair. Manuals
 I. Title

ISBN 0 7506 0102 7

Printed and bound in Great Britain by
Butler & Tanner Ltd, Frome and London

CONTENTS

PREFACE

It is many years since an attempt was made to write a truly practical book on servicing domestic electronic products, whose levels of sophistication are now very high.

The emphasis in this one is on modern and recent set design, but recognition is made of the fact that the majority of equipment which finds its way to the repair bench is 'middle-aged'. A place has been found, then, for thyristor-operated TV designs, for top-loading VCRs and for multi-chip colour decoders amongst the cut-dried-and-sealed microchip technology of today's products.

No book can do the repair job for you! Correct test equipment, good training and adequate theoretical knowledge are essential. To help with the latter two, my books *Newnes Guide to TV and Video Technology* and *TV and Video Engineers Pocket Book* look at the same areas from a more theoretical point of view.

The task of the serviceman is seldom an easy one. If this book smoothes his path, and turns round repair jobs more quickly, it will have achieved its aim.

My thanks are due to colleagues Hugh Lobo and Terry Skilton for their suggestions at the outset; and to my wife Anne who turned 450 pages of scrappy handwriting into a typescript while looking after a lively family.

E. Trundle

1

INTRODUCTION

Domestic TV and video equipment has become very sophisticated in recent years, and the pace of development shows no sign of slackening. While the number of engineers employed in repairing and servicing 'brown goods' has decreased, the demands made on those who are left is steadily increasing. The trend is away from 'stock faults' and diagnosis-by-substitution, since products have become more reliable and diverse in type, components and design. More than ever, it is required of the engineer that he repair, service or set up equipment which he has seldom or never seen before—if he is very familiar with the product the implication is that it is unreliable!

To be successful in this environment the engineer needs more aids than were necessary in days gone by, when models were changed less often, when circuit and mechanical design varied little between manufacturers, and when a large proportion of the faults in individual models were due to known weak spots or wear-prone components. Here true diagnostic ability was not necessarily the main requirement. Now the situation has changed to the point where most servicing has to be carried out in a workshop environment, aided by sophisticated test equipment and calling for skills and training of a high order. This gives rise to an anomaly between the purchase price of mass-produced hi-tech equipment and the cost of servicing and repairing it; the effects, apart from sometimes souring the customer's attitude to the serviceman, are to put rigid constraints on what is economically viable in the repair field; and a tendency for products to be scrapped and replaced after a relatively short life-span.

The key, then, to effective and profitable servicing is rapid and accurate diagnosis of the fault. Knowing *exactly* what is wrong with a piece of equipment is but a short step from having it working again, and the most successful engineers are those who have the ability to quickly track down, to component level, the root of the problem. To achieve this, many ingredients are necessary:

1 A good working knowledge of the theory of operation of TV and video equipment, which can only come from study – at a technical college, from textbooks (i.e. *Newnes Television and Video Engineer's Pocket Book*) or from an accumulated bank of experience.

2 A specific knowledge of the model to be dealt with, ideally from a manufacturer's service training course, but often of necessity from the service manual and any circuit/operation description it contains.

For all but the simplest repairs a service manual is essential, and it is foolhardy (and time-wasting!) to attempt diagnosis and repair without it.

3 A range of suitable test equipment – and in the case of VCRs and cameras, jigs – to act as eyes, ears and rulers for the engineer. The next chapter will describe the characteristics and applications of test gear, and outline what is required for various categories of servicing.

4 An environment in which you can concentrate and be comfortable. In practice, this means minimizing interruptions by people and phone calls; large, well laid-out benches with the work area separated from reception and 'transport' departments; good lighting, both overall and local; good heating and ventilation, and insulation, where necessary, from ambient noise.

5 Good lines of communication with the manufacturer or his agent. This is important for the supply of service manuals and modification data; the exchange of technical information; and an easy flow of spares. Many manufacturers do not deal at all with other than their appointed agents and service centres for technical advice, parts supply and training courses.

6 Concise and specific information on the fault symptoms from the owner or user of the equipment. This is especially important where the problem is a borderline or intermittent one.

1

Many of the factors outlined above are responsible for the trend in the service industry towards 'service centres' where test equipment, expertise, service manuals and other resources can be pooled in a 'group practice' wherein the economies of scale come into play. Here the throughput of repair jobs can be made high enough to justify the tooling and running costs. An alternative approach is the 'specialist' one, where an individual or group of engineers concentrate their talents and resources on a limited range or make of equipment, and achieve sufficient throughput by drawing on a wide geographical area.

Many new recruits to the trade, and a good many established engineers, especially when faced with new or unfamiliar models and techniques, have difficulty in diagnosing faults due to what might be called the 'jungle' effect. When faced with a fault the reaction can be 'where do I start?' or 'I wish I'd never picked this one up!' The purpose of this book is to guide its reader as quickly as possible to the fault area and, hopefully, to the specific component or maladjustment which is responsible.

A TV, VCR or camcorder contains many separate sections and there is a great deal of interdependence between them. With this and the increasing use of digital techniques in control and signal-processing stages, the symptom does not always provide a strong indication of the fault area as in the past. In these cases the correct fault-finding approach can save precious and expensive time. It is not possible in a wide-ranging book like this to provide specific information on individual circuits, makes and models. The block diagrams of all makes and models of equipment doing the same job to the same standards, formats and principles are almost identical, however, and even within the 'blocks' the circuitry and IC functions are surprisingly similar in many cases. Throughout the book, then, we shall be looking primarily at block diagrams and the key interconnections between them. As an early step in the diagnostic procedure the study of the manufacturer's overall block diagram, and those of the individual sections of the TV, VCR or whatever is very important in narrowing down the field of search. Once the faulty block is identified, individual sections and components within it can be investigated, starting with a check of the operating voltage, then following the sequence: signal in, signal out, control influences, operation/distortion. Where, as is often the case, the components and functions within the block are common, we shall look into the block in detail.

The first step in the diagnostic process is a careful and thorough analysis of the symptoms: as displayed on the TV screen or indicated on suitably-applied test equipment. The importance of this cannot be over-emphasized. A few minutes spent thinking about the symptom and its possible cause (especially where more than one symptom is present) can save much time in fruitless investigation of innocent sections and components. Some examples illustrate the point: a noisy, snowy off-air picture in a TV set may be due to a faulty aerial system, quickly checked with a field strength meter or another receiver before dismantling the TV; and if the fault *is* in the set, it's unlikely to be much further downstream than the tuner itself if the picture-noise level is high. Similarly, an unlocked servo in the VCR may be approached by an in-depth check of the servo circuit, only to find that the control-track head is dirty!

Visual examination

Once it's proved that the fault is in the equipment and not in such peripherals as aerial, monitor TV, connecting leads, tapes, batteries etc. it's a good idea to make a close visual inspection of the appropriate part of the equipment, be it a printed-circuit board, a tape deck, a fluorescent display panel or a picture tube. Many faults are due to physical causes, amongst which are cracked PCB tracks; dry joints; liquid ingress and subsequent corrosion; broken components; foreign bodies causing electrical short-circuits or jamming tape mechanisms; lost vacuum in thermionic display devices; corrosion from neglected batteries or from decomposing electrolytic capacitors; and so on.

With picture-tubes, a close inspection of the equipment operating in darkness will reveal any corona discharge, sparking or tracking not otherwise visible – and whether the heater is alight! In relatively high-current circuits arcing at soldered joints, plug/socket connections and switches can more easily be seen in darkness.

Visual inspection of a malfunctioning VCR deck can also save much time. Many picture, sound and 'function' faults are due to slipping or broken drive belts, incorrect tape path, stiff mechanics, broken pinion teeth, failure of a member to rotate, the development of tape slack, or similar causes, all to be dealt with at length later in the book. Suffice it here to say that time spent looking at and listening to decks is seldom wasted.

Apart from detecting actual fault causes, 'circumstantial' evidence and useful clues to the fault area or component can be gleaned by careful examination. A discoloured resistor may well have overloaded and changed value; a pile of powdered or shredded tape debris at a certain point on a VCR deck is a useful indication of where the tape is being damaged; much can be surmised about why a fuse failed by looking at its dead body; score-marks on moving deck members reveal excessive friction, wear and damage – many more examples will be given in the following chapters.

Even the sense of smell plays a part in the diagnostic process! An acrid, dust-burning smell from high-rated resistors indicate that equipment has been a long time out of use; the distinctive smell from overheating resistors, faulty EHT multipliers, ozone from corona discharge, and overheating varnish on wound components are distinctive and unmistakable to the experienced engineer, leading him quickly to the root of the trouble. The unpleasant smell of animal urine or flower-pot water warns him to expect contamination and corrosion of mechanical and electrical components, and points him in the direction of horizontal surfaces beneath ventilation slots.

Service process

As a rule, a repair job has four distinct aspects: diagnosis; repair (usually replacement of faulty components); setting up/alignment; and testing. It is important to separate these as far as possible, though sometimes they necessarily run together, as when diagnosis has to be done by trial-and-error subsitution of components, and when the setting-up process consists of adjusting or checking for a specific performance level. The most crucial and difficult of these processes is diagnosis. The physical aspects of repair and component replacement are covered in Chapter 21, while alignment and testing, though touched upon several times in this book, is the province of the service manual and specification table of the equipment manufacturer.

In general, throughout the book we shall be referring to colour equipment. In the few instances where monochrome signals only are handled, only the information on chrominance stages need be disregarded – all the other sections of the TV, monitor, camera, VCR etc. follow closely the configuration given. In practice black-and-white TV sets and monitors make smaller current demands on their timebase and power supply sections.

Diagnostic short-cuts

To make a full step-by-step diagnosis can be time consuming, and it's often possible to bypass several stages in the checking process. Long experience has established that the following categories of components are most vulnerable to breakdown:

1 High-value resistors (i.e. > 100kΩ), especially when subjected to high voltages, as in picture-tube, power-supply and some timebase applications.
2 Electrolytic capacitors, which being chemical devices, have a finite lifespan. Old electrolytics tend to 'dry up' and lose capacitance. Some failures are detectable by inspection, when corrosion or white deposits are evident around the leadout wires. Capacitors in warm surroundings, and those passing a heavy ripple current are most prone to failure.
3 Electrical components where connection is maintained by spring pressure, i.e. switches, variable potentiometers and plug/socket connections of all types. These can give rise to intermittent faults due to oxidization or loss of pressure as a result of metal fatigue.
4 Soldered joints which pass a heavy current and/or physically support a heavy component or one which is subject to mechanical stresses. Those associated with chopper- and line output-transformers, with mains on-off switches and with camera- and remote control sockets fall into this category.

Where the cause of a fault is not clear, typically a no-colour situation in a TV or VCR or poor 'trick' replay of a video tape, useful clues to the culprit can be gleaned by following the manufacturer's alignment or setting-up process. If this does not itself clear the problem, it will generally show what is *not* happening, or what has run out of road, as it were.

Checking transistor junctions

A rough check of suspect transistors can be made with them still in circuit. While a special transistor tester can be used, an ordinary multimeter is usually adequate. The transistor, when correctly polarized, should conduct from base to emitter and base to collector, but not vice versa, and not at all between collector and emitter in either direction. A voltage drop of 500–800mV should be seen across each conducting junction – bear in mind that many multimeters on resistance ranges produce a positive potential from the negative (black) test prod.

The same test applies to diodes, and in both cases it is sometimes necessary to disconnect the device from circuit to avoid the shunting effect of other components. This is most easily done by disconnecting all but one leg (for physical support) from the printed-circuit board.

Meter checks

The small test voltage/current available from analogue meters such as the Avo models 8, 9 etc. in resistance test ranges can be put to other uses. It will turn on a semiconductor junction, and on the lowest ohms range can run many small d.c. electric motors at low speed – a useful test. When switched to a.c. or d.c. voltage ranges, various known resistances are present across the prods, useful as a quick substitution value in circuits where the current will not overload the meter movement. With the meter switched to its highest current range a handy and easily applied shorting-link is available across the probes.

Empirical tests

In attempting to find the cause of a problem there is a great temptation to twiddle pre-set controls to see what effect they have, and whether they have any bearing on the fault. This can be useful (and justified) where the pre-set is a line hold control in a TV whose line speed is wrong, or a 'set voltage' control in a PSU whose output level is low. Random twiddling is dangerous, however, and unless you have an understanding of what the control should do, and the means (i.e. setting-up instructions and suitable test-gear/jigs) of correctly resetting it, do not twiddle any pre-set without first very accurately marking its original position.

Connecting test equipment

Making and maintaining contact with test-probes of meter, counter or oscilloscope is a difficult

business on PCB connections – carelessness can cause a damaging short-circuit. The problem is simply overcome by heating the print land or connection and melting on a short stub of solder to act as connection-post. Don't forget to remove it when tests are complete.

Test cards

Test cards and patterns are important tools in checking performance, setting-up and alignment of video and TV equipment. For many years now they have been generated or stored electronically, and come either off-air from the broadcasters or cable-head; or from a pattern generator in the workshop. Patterns devolve into two types, specialist ones used for checking or adjusting one aspect of TV or VCR performance (to be dealt with in the next chapter), and general ones containing many components for overall checking of a system and its display unit.

Two typical electronic test patterns are shown in Figures 1.1. (ETP1, designed by the IBA) and 1·2 (PM5544, designed by Philips).

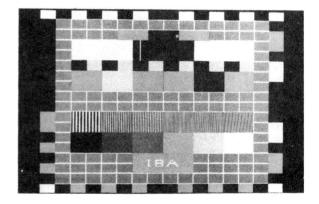

Figure 1.1 *ETP1, the IBA's broadcast test card*

ETP1 test card

This general-purpose card is used by the IBA during non-programme hours, and contains features to check most aspects of TV and VCR performance:

1 *Border castellations to mark the picture limits.* They should be at least one-half visible on all four sides; aspect ratio is 4:3. The top blocks are cyan

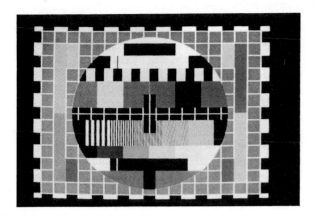

Figure 1.2 *Philips PM5544 test card*

to check the performance of the chroma reference generator after the field blanking period, during which the colour synchronizing bursts are absent. Those on the left-hand side are red (upper) and blue (lower) to check burst-gate timing in the receiver's decoder; late pulse-timing causes hue errors in horizontal bands aligned with the border blocks.

The right-hand (white and yellow) and bottom (green) castellation blocks check chroma reference generator performance at the end of line and field periods respectively.

2 *Crosshatch pattern.* The white grid on a grey background gives a check on colour registration (convergence) between the three coloured rasters which make up the picture. There should be no colour fringing at screen centre, and little more than 1mm at the edges, 2mm at extreme corners.

3 *Colour bars.* One-third of the way down the picture, these blocks contain EBU standard colour bars at 75% amplitude, 100% saturation. Left-to-right they are yellow, cyan, green, magenta, red, blue and black. While they form a useful visual check, decoder testing and setting-up is better done on a locally generated full-screen colour bar pattern (see Chapter 2) which is easier to synchronize and analyse on an oscilloscope.

4 *Grey-scale step-wedge.* Two-thirds of the way down the card come six blocks of progressively brighter grey, ranging from black to full white, 0/20/40/60/80/100% amplitude of the luminance signal. Brightness is correct when the black block (LHS) is *just* extinguished; approximately equal increments of brightness should now be visible between adjacent blocks.

5 *Multiburst.* Above the step-wedge are blocks containing progressively finer vertical black lines, corresponding to luminance frequencies of (left to right) 1.5, 2.5, 3.5, 4.0, 4.5, 5.25MHz. Only high-definition (e.g. monochrome monitor) sets can resolve all six, which form a severe test of resolution, focus, alignment, and vision channel bandwidth. A typical domestic colour TV and an S-VHS video will resolve the first four, a standard VHS video the first two. In most systems, unresolved blocks are affected by cross-colour patterning.

6 *Black/white rectangles.* About a quarter of the way down the picture is a broad bar containing white/black/white. Its purpose is to check the low-frequency response of the luminance channel. Any graduation in brightness across the width of each block, or smearing from right-hand edges indicates poor l.f. response. The white needle in the black rectangle shows signal reflections and consequent 'ghost' images to its right. With experience, this and the rectangles' appearance can be used to judge the performance of a VCR's edge-sharpening (enhancer) circuits.

7 *Black/grey squarewave.* Below the black/white rectangles are twelve blocks, alternately black and light grey. These check transient and m.f. video response, and should have well-defined edges. Within the blocks there should be no streaking, smearing or 'overshoot'.

8 *Colour registration block.* Near the top of the test card is a red bar sandwiched between two yellow ones. Incorrect timing between luminance and chrominance information is shown by colour 'bleeding' between red and yellow. Misregistration in a video-playback pattern is often caused by incorrect tape back-tension.

9 *Luminance level.* The overall brightness of the test card corresponds to a well-balanced normal colour picture.

Philips PM 5544 card

Figure 1.2 shows a widely-used electronic test pattern, which contains many similar features to ETP1. The extra features in PM5544 are as follows:

1 *Circle.* Useful for checking horizontal and vertical linearity, where errors readily show as 'egg' or 'potato' shapes.

2 *Colour-pattern 'brackets'.* On each side of the central circle, these give comprehensive

information on the alignment and performance of a colour decoder. The left-hand one contains − (R-Y) in its upper half, and + (R-Y) in its lower half. The right-hand one has − (B-Y) in its upper half and + (B-Y) in its lower half. The rectangles which form the top and bottom corners of the brackets contain colour information at phasers 326° (top two) and 146° (bottom two), both points corresponding to zero G-Y information.

In the space between each 'bracket' and the edge castellations is transmitted 'colourless information' in the form of unswitched R-Y (LHS) and *switched* B-Y (RHS). Correct decoding cancels out these signals to give a colourless strip indistinguishable from the rest of the pattern background. If a decoder phase error is present, however, off-axis demodulation will take place to give a blue or green tint on the left strip and a pink or yellow tint on the right strip. Similarly, errors in PAL delay-line amplitude and phase settings give rise to Hanover bars in the main sections of the brackets.

3 *Interlace check.* Correct interlace is indicated when the centre horizontal crosshatch line is equal in thickness to the others on the card. A difference here indicates incorrect or erratic firing of the field timebase, and should direct attention to the field sync triggering section.

BBC test card G

This pattern is based on the PM5544 configuration described above. The BBC version has 95% saturation, 100% amplitude colour bars and sinusoidal-waveform frequency gratings corresponding to 1.5, 2.5, 3.5, 4, 4.5 and 5.25MHz.

There are other, less significant differences.

Block diagrams

Throughout the book the separate sections of TV and video equipment are dealt with individually. In diagnosis work, unless you are very familiar with the product on the bench, it is often necessary to start with the manufacturer's block diagram for the whole set, and progressively narrow the field of search. Each of the main blocks devolves into separate block diagrams, which will here be dealt with in the appropriate chapters. Many of the 'secondary' block diagrams in fact consist of ICs,

which are d.c.-coupled internally. The 'block' approach to servicing is an important key to rapid diagnosis, and (particularly with ICs) can avoid time-wasting 'red herrings'. Unless the d.c. input conditions (and very often the pulse feeds) are correct on an IC the output signals and d.c. levels will certainly be wrong; and many an IC has been changed, only to find the symptom unchanged and the chip's input conditions wrong.

To tie together the separate chapters, overall block diagrams are given here, with individual blocks annotated with corresponding chapter numbers. Not all equipments will contain all the blocks listed, and there is greater interdependence between sections than can be shown here. Similar symptoms can arise from quite different sections in some cases; to cater for this there is a *symptom index* at the back of the book.

Once the fault area has been located, the sub-block diagram and circuit diagram in the service manual can be consulted. Many manufacturers, to avoid a complex maze of lines on circuit diagrams, use a trunk-and-branch routing system with entry and exit points for individual lines marked by codes and cyphers. It is essential to master the coding system (key given in manuals) in order to make sense of the diagram!

TV block diagram

Figure 1.3 shows a typical block diagram of a TV set, with key waveforms. The UHF signal enters the tuner where it beats against a stable local oscillator to translate it to a fixed i.f. frequency. The bandpass response is defined in a filter then the signal level is raised in a multi-stage i.f. amplifier under the influence of a.g.c. control. The signal is

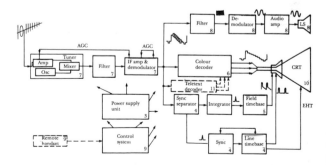

Figure 1.3 *Functional block diagram of a TV receiver. The numbers correspond to the chapters of the book*

demodulated and passed to a decoder section whose RGB outputs are amplified to a high level for presentation to the picture-tube cathodes. The audio signal appears in the form of a 'beat' output from the vision demodulator at 6MHz, to be filtered off, amplified, clipped and f.m. demodulated to baseband. It is subsequently amplified and passed to the loudspeaker in mono TV sets; decoded into separate L & R channels for their own amplifiers and speakers in stereo models.

The sync separator strips the sync pulses from the video signal; separate filters process the line and field sync pulses for timebase triggering. The triggered timebase oscillators define the timing of field and line sawtooth waveforms driven through the scan yoke by their respective output stages. The line output stage is usually the source of various 'secondary' supplies in the TV such as operating voltages for the picture tube and for some of the other operational blocks shown. It – or the line oscillator – is also the origin of various gating, clamping and keying pulses distributed around the TV. The power supply section (sometimes integrated with the line output stage) provides primary power and operating voltages for all the blocks, directly or indirectly.

'Optional' sections of the TV include the electronic control system, which interfaces the user with the operational parts of the set such as tuning, analogue controls and teletext command and selection. The teletext section takes a video signal from the vision detector and gates out the teletext data in the field blanking interval. Selected data is decoded and used to direct a character generator whose outputs, in RGB form, are applied to the RGB video amplifiers.

VCR block diagram

A functional block diagram for a homebase VCR is given in Figure 1.4. The receiver section, up to the sound and vision demodulators, is similar to that of a TV. The post-detector sound processing depends on the type of machine. In conventional types the signal is merely added to an a.c. bias signal and applied to the audio recording head; the audio bias oscillator also supplies the *all-erase* head which wipes clean the tape in preparation for recording. In Hi-Fi machines a great deal of processing is carried out on the audio signal before its application to the rotary heads on the head-drum. The replay processing is the inverse to that carried out on record.

The vision signal is first filtered into chrominance and luminance components. The luminance part is a.g.c. controlled then clamped, pre-emphasized and clipped for application to a VCO, whose output

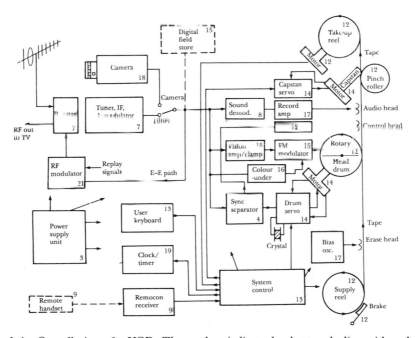

Figure 1.4 *Overall view of a VCR. The numbers indicate the chapters dealing with each section*

(an f.m.-modulated carrier) is applied via a rotary transformer to the video recording heads. The separated chrominance signal is down-converted by a heterodyne process to a frequency around 700kHz before being added to the luminance signal for recording.

The broadcast field sync pulses are stripped from the luminance signal and used in the head-drum servo to synchronize head sweeps to TV field scans. Simultaneously a frame-marker pulse is laid on the tape at 40ms intervals by the *control-track head* for phase-lock purposes during playback. The other main servo, that for the capstan (the prime mover of the tape) maintains a steady tape speed through the deck.

The operation of the deck and the electrical sections of the VCR is controlled and co-ordinated by the system-control (*syscon*) section, working from the user's key-pad and from feedback signals from deck-sensors. The syscon also interfaces with the clock/display section – the user's 'status-readout'.

During playback the same steady tape speed is maintained by the capstan servo to permit orderly readout of the recorded vision and sound signals. The alignment of head sweeps with tape tracks is ensured by the servo system, which now compares head-tacho pulses with off-tape control-track pulses to steer either the head drum or capstan to the correct phase-point, adjustable with the tracking control.

The off-tape signals are sorted apart by bandpass filters. The high-frequency luminance f.m. carrier is amplitude-limited and has its dropouts patched before application to a demodulator, whose output, after filtering and de-emphasis, forms the baseband output signal. It is added to the off-tape chrominance signal which has now been up-converted and treated to remove timing jitter. This reconstituted composite video signal now passes out of the machine, in baseband form at the video output socket, and modulated onto an r.f. carrier in the machine's *r.f. convertor* secion. At this point, too, the vision signal is reunited with the sound signal, the latter having come off-tape via an f.m. demodulator (Hi-Fi) or equalizing amplifier (conventional).

The electrics and mechanics, directed by the syscon, are energized by the power-supply section, which in table-top VCRs is a converter and regulator of domestic-mains energy. In VCRs various types of internal supply lines are necessary: closely regulated low-energy ones to operate the signal-processing and logic (i.e. syscon) sections; and high-energy lines to power motors, solenoids etc.

All the sections illustrated in Figures 1.3 and 1.4 are described at length in the appropriate chapters from the point of view of fault-finding and diagnosis.

Fault-finding trees

For use from a 'cold-start', Figures 1.5 (TV) and 1.6 (VCR) give some indication of the overall fault-finding process, which initially consists of tying the trouble down to a single stage within the machine.

Whenever the equipment comes in the form of separates, i.e. camera/recorder/tuner-timer/power unit, etc. it is best policy to have all equipment on

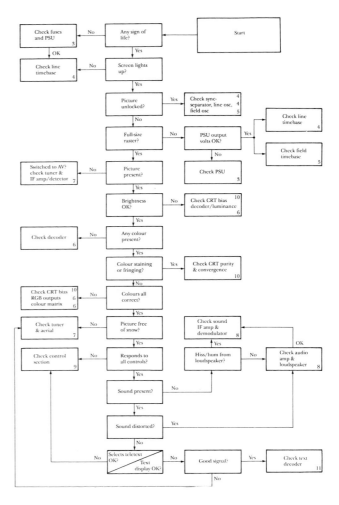

Figure 1.5 *Basic fault-finding procedure for TV. The 'end-box' numbers refer to the chapters in the book. See also the symptom index on page 207*

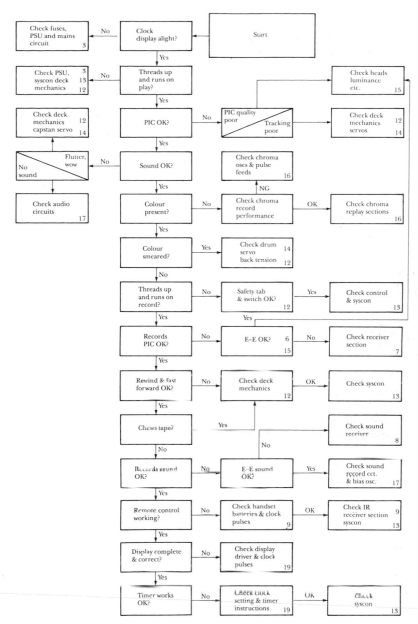

Figure 1.6 *Direction-finder for VCR troubleshooting. The 'end-box' numbers refer to the relevant chapter. See also the symptom index on page 207*

the bench together and treat it as a complete ensemble for test-and fault-diagnosis purposes. This ensures that the fault will show up, wherever it lies, and saves the trouble and time of providing and interfacing signals, power supplies etc. For the same reasons, and especially where a record problem is intermittent or a tape-damage fault is reported, a sample cassette is also very useful.

Repair viability

A typical lifespan for TV and video equipment is 5–8 years; much depends on the hours of use it sees. It is important, when diagnosing faults and estimating costs, to bear in mind that if the equipment is approaching end-of-life, or in special circumstances like power-surge or lightning damage, more than one major component may have failed, a situation only revealed when the first is replaced at considerable expense. Take the precautions of testing the picture-tube emission (see Chapter 2) of a well-used TV whose line output transformer has failed, and of checking the video heads of a VCR whose motor (or other expensive part) is in need of replacement. This 'hidden fault' situation accounts for the difficulty of accurately quoting or estimating for repairs, and for the fact that more often than not the repair has to be at least partially done to gauge the depth of the trouble and arrive at an accurate figure for repair cost. Figure 1.7 shows an alternative to repair for old and obsolescent sets. . . .

Figure 1.7 *The end of the road for life-expired equipment?*

Preventing 'bounce-backs'

One of the problems of the TV/video service business is the 'bounce' job, where the equipment is returned for further service after repair, with a problem that may or may not be associated with the previous work, but which is always expected to be rectified free of charge. Unlike a motor car or a central heating system, whose separate components and their functions are visible and obvious, electronic equipment tends to be regarded as a single entity by its owner or user, who would almost certainly blame the repairer of a no-picture fault for a no-sound fault developing shortly afterwards; but would not associate the failure of their car's headlamp bulb with the recent fitting of a tyre or an exhaust silencer!

To minimize this sort of problem, ensure that newly soldered joints are well made; that old joints are not crystallized or 'dry'; that internal leads are properly dressed; that mains plug connections are properly made; that electrolytic capacitors are not visibly deteriorating; that no arcing, sparking or corona discharge is present; and that known trouble spots for that model – i.e. belts, idlers, motors, electrolytic capacitors, switches etc. – are checked. It is also important to make clear to the owner or user at the outset or at the time of quoting or estimating costs that faults arising in areas other than that repaired cannot qualify for free or subsidized service after the job has been done.

Safety

In servicing electronic equipment safety is of paramount importance. The main areas concerned are the safety of the serviceman while he is doing the job; and the safety of the equipment in use afterwards, from the point of view of electric-shock, implosion and fire hazards to the user. These aspects of servicing are covered in Appendix 1, reference to which is recommended before completing and handing over repaired products.

2

TEST AND DIAGNOSTIC EQUIPMENT

Good test equipment is absolutely essential to the successful and efficient servicing of TV and video equipment. Only the simplest and most obvious faults can be dealt with without test gear; and since the meter, oscilloscope, etc. are the only interface between the engineer and the equipment, it is vital that they are accurate and reliable – and that they are well looked after! The most important instruments for general practice are the oscilloscope and the multimeter, and it is with these two that most diagnostic work is done.

As will become clear, however, there are many other classes of test equipment. Some are essential for serious service work: a pattern generator, an isolating transformer and a picture-tube tester/reactivator for TV servicing; a frequency-counter, various test-tapes and mechanical jigs for VCR testing and repair. Some test equipments, like signal strength meters, satellite dish alignment aids and vectorscopes are only applicable to certain products and engineer-roles; the usefulness of others, like component test-bridges, signal injectors and reference light sources depends entirely on the attitude and working method of the individual engineer. Most *test-jigs* are specific to one make, model or range of equipment; and their acquisition, if expensive, can only be justified where the service throughput of that product is high. As a rule, it is wise to stick to the products for which you are equipped.

In general, test equipment is good value for money, and will recoup its cost in a reasonable period. It is important, though, to choose equipment that is appropriate to the job in hand and to steer a course between cheap, cheerful and flimsy gear whose accuracy, reliability or robustness is poor; and spending more than is justified on equipment whose full potential will never be used, or which has such an obscure function that it spends most of its time sitting on the shelf.

In order of importance the equipment needed for TV fault diagnosis is a dual-beam oscilloscope with probes; multimeter; isolating transformer; variac;

degaussing coil or wand; tube tester; test pattern generator; variable low-voltage PSU; and logic probe. For VCR fault diagnosis and setting-up: an oscilloscope; multimeter; alignment tape; back-tension gauge; torque gauge; frequency counter; test pattern generator; logic probe/analyser; variable PSU; and the various mechanical jigs and fixtures specified and supplied by the VCR manufacturer. These 'mechanical' devices will be dealt with in Chapter 12, and hand tools in Chapter 21. For the rest of this chapter we shall examine the types, specifications and roles of the main classes of test equipment used in TV and VCR servicing.

Oscilloscope

The oscilloscope is the most useful and versatile instrument in the repair workshop: at a pinch it can replace the multimeter and frequency counter, as well as playing its own unique role of waveform display. Essentially the oscilloscope has a screen upon which a spot of light is deflected against a calibrated graticule. Horizontal deflection is generally by means of an internal *timebase* generator, whose ramp output waveform is linear to give an adjustable timescale across the screen from left to right. The start of each sweep can be initiated (*trigger* function) from a feature of the waveform displayed, or from an 'external' pulse at the choice of the user. The vertical position of the light-spot is governed by the signal in the Y-amplifier(s) to give a continuous plot of voltage against time. Figure 2.1 shows a photo of a good general-purpose service oscilloscope.

CRT

The tube used in a scope, and its operating conditions, determine the instrument's price, class and performance. For our purposes, a tube with an internal graticule is preferable, working with as

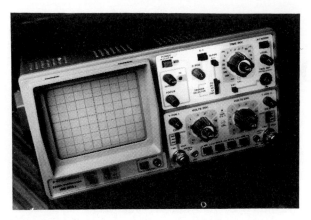

Figure 2.1 *A bench oscilloscope suitable for TV and video work by Hameg*

high a PDA (post-deflection acceleration) potential as can be reasonably afforded. Standard scopes have 2kV EHT potential, whose brightness is adequate for most purposes. For very low duty-cycle displays (like examining teletext data lines and small portions of video f.m. envelope in VCR replay) a higher PDA voltage is required for a reasonably bright and well focused trace. Regarding size, a ruled area of 8 × 10cm is a reasonable compromise between price and practicability. The green medium-persistence phosphor type P31 (Jedec), or GH (Pro-electron) is most suitable for general service work.

To prolong tube (and hence instrument) life, do not leave the scope running continuously throughout the working day unless it is in constant use. Modern instruments warm up and stabilize very quickly.

Y-amplifier

The two most important characteristics of the Y-amplifier are bandwidth and sensitivity. Virtually all the signals of interest in TV and video servicing are below 10MHz. Even so, a 20MHz or 25MHz scope costs little more (in relative terms) than a 10MHz type and is worth having for its better rise time and other characteristics.

Sensitivity-wise, performance has improved in recent times, and 5mV/div instruments are now commonly available, and should be regarded as a minimum standard. Better if it can be afforded, is a 2mV or 1mV type, especially since the scope is generally used with a 10:1 probe (see below) which desensitizes the Y-amplifier by a factor of ten. Some wideband oscilloscopes have a Y-magnification facility which increases Y-gain – typically by a factor of five – at the expense of bandwidth. Y-gain is normally set by a multi-step attenuator with 2-5-10 sequence: a top setting of 20V/div permits examination (via the 10:1 probe) of flyback pulses at the collector of a line output transistor. A *vernier* control provides gain adjustment between switched ranges, but must be left in the click-stop position to validate the calibrated settings of the gain switch. For operating convenience, the Y-amplifier is normally a.c.-coupled by a large capacitor at the input terminal: this permits easy examination of a small a.c. waveform riding on a large d.c. potential. Other settings of the input switch offer d.c.-coupling (facilitating the scope's use as a fast-acting d.c. voltmeter capable of positive and negative readings); and *ground* which permits a reference base-line to be set with the Y-shift control.

Dual-trace capability is a 'must' for service work. This involves two separate, identical Y-amplifiers whose outputs are alternatively switched to the Y-deflection plates of the CRT: on alternate horizontal sweeps at high X-scanning speeds, and on a high-speed 'chop' basis at low X-scanning speeds. Dual-trace capability permits simultaneous display of two waveforms for phase/timing comparison. True dual-*beam* instruments do exist, but at prices beyond the reasonable reach of a TV/video service department!

Timebase section

The sweep generator offers a range of speeds appropriate to the capabilities of the Y-amplifier. A typical 20MHz scope may have a fastest setting of 200ns/div, while a 60MHz instrument would offer maximum sweep speed of around 50ns/div, both usually expandable by a × 5 or × 10 'mag' facility. To fit a horizontal feature exactly between graticule lines if required a vernier speed control is provided, again with a click-stop setting to validate the main switch settings. The sawtooth output from the sweep generator is often brought to an external socket for use with a *wobbulator* or similar instrument.

Many oscilloscopes have a facility for external input to the horizontal deflection amplifier (X-Y mode) for the display of Lissajous patterns. While this is seldom required for TV or VCR servicing, it is often used for setting up audio CD players – a point worth bearing in mind when selecting an oscilloscope.

Trigger section

Virtually all the waveforms of interest to us are continuous or repetitive, and so long as the sweep is triggered at the same point in the cycle the images traced out will overlay. The job of the trigger section is to select a particular feature of the Y (or external) signal and apply it to initiate the sweep. Trigger selection can be made between positive and negative peaks; a.c., d.c. and h.f. coupling, the latter via a high-pass filter; from the line or field-rate outputs of a built-in TV sync separator; or from 50Hz mains. Some oscilloscopes have a useful line-pulse counting trigger circuit, permitting the sweep to begin on the same TV line in each frame. A useful feature is the provision of a short delay line in the Y-channel(s) to permit display of the event that triggered the sweep.

Very often there is no feature of the Y-signal to reliably trigger the sweep; examples are pure chroma signal and VCR f.m. replay envelopes. In these cases external triggering (from line pulses and head flip-flop pulses respectively) is necessary.

Probes

To prevent excessive loading on the circuit under test a 10:1 probe is generally used – it brings the input impedance up to 10MΩ and 10pF. To cater for different oscilloscope input capacitances the probe incorporates a trimmer, whose correct setting is essential for correct readings at high frequencies. To set, apply a squarewave of about 1kHz with fast rise time (often provided at the front panel of the scope) and adjust for square corners – see Figure 2.2.

(a)   (b)

Figure 2.2 *Setting up an oscilloscope probe: adjust the trimmer to avoid the spikes at* (a) *and the rounded corners at* (b)

Multimeter

A general service multimeter has ranges for a.c. and d.c. voltage and current up to about 1000V and 10A, together with resistance ranges to read between say 1Ω and 20MΩ. The two basic forms of meter, analogue and digital, both have pros and cons; buying a cheap meter is often a false economy.

Analogue meters

The traditional analogue meter, with dial and pointer, has served the trade for many years, and many engineers still prefer them. The passive types typically have a circuit-loading resistance of 20kΩ/V on d.c. volts ranges, less on a.c. ranges. Their accuracy is around ± 2½% and they are somewhat fragile, electrically and mechanically. They do give a 'feel' for the circuit under test, however, and the 'kick' of the needle on resistance test of a capacitor gives a good idea of its capacitance. The high applied voltage and relatively high test current on high and low resistance ranges respectively also gives more versatility when testing semiconductors like zener diodes and high current devices. The 'resistance between prods' tricks described in Chapter 1 are also unique to passive analogue multimeters. For checking *relative* values of voltage and current, decibel scales are useful.

Ideally both analogue and digital meters should be available! In choosing an analogue meter, aim for 20kΩ/V d.c. and especially for an effective and reliable overload cut-out circuit which offers protection to the internal resistors as well as to the meter movement itself.

Digital meters

Multimeters with digital readout are much more accurate (typically ± 0.5%) than analogue types, and for the same intrinsic quality are cheaper. They have the advantages of high input impedance (i.e. 10MΩ); an ability to work at any angle or situation; a mechanical ruggedness which no analogue meter can match; and an immunity to electrical overload only breached by *very* high d.c. or pulse voltages. The DVM's voltage and current ranges also reach lower than those of a passive analogue meter.

For general service work, even on the bench, a battery operated instrument is more convenient than a mains-powered type; the LCD display is very economical of battery power. There is little point in spending a lot of money on a highly accurate 4½ digit instrument when a 3½ digit type with 1% or 0.5% accuracy is perfectly adequate. Most important, and this applies to all test equipment, is to ensure that spares and service will be available throughout the instrument's life, and this means buying the product of a creditable and reputable company.

A probe for checking EHT and focus voltage is sometimes required. No very great accuracy is needed here and a separate 'wand' or stand-alone type is perhaps more convenient than a probe for use with a multimeter or scope.

Picture-tube tester/reactivator

This instrument powers up the picture tube with the required heater, cathode, grid and accelerating voltages, then measures beam current as an indication of the emissive capability of the cathode – separately for each beam in the case of a colour tube. The emission is an important indicator of the age, condition and state of wear of the picture tube. In general when a picture tube is worn out the set is scrapped owing to the high cost (in terms of that of a new set) of a replacement tube and the labour in fitting it.

Most cases of low emission in picture tubes are due to oxide coatings on the emissive surface of the cathode, a form of 'poisoning'. The coating can be stripped to reveal a new and clean surface, which gives the tube a new lease of life. This reactivation is carried out by running the cathode at a higher-than-normal temperature while passing relatively heavy current pulses from it to the control grid. The process has to be carefully controlled to be successful, and to prevent damage to the cathode under treatment. Some tubes are more amenable than others to this process, and reactivation is not possible in any case when the cathode surface is exhausted – or where some other problem is present like O/C heaters or electrodes; or shadow-mask or purity defects. Most tube tester/reactivators have one other 'repair' process available: a 'clear-leak' facility whereby a large capacitor is discharged through any conductive flake or particle which is lodged between electrodes. The heavy current can 'explode' and disperse the offender.

Tube testers come with a range of base sockets to suit the various tube types encountered. In practice the best instruments are those with three meters on which the emission and warm-up times of the three guns can be simultaneously checked; and two or

more levels of reactivation pulse current for different circumstances.

Variable power supply

In TV and video servicing there is often the need for an independent source of voltage or current to assist in diagnosis and setting-up. Sometimes all that is required is a low-current bias potential to give a reference voltage or to drive an a.g.c., servo or varicap tuning system to a required point: these requirements can often be met by a battery and potentiometer, sometimes referred to as a bias box. The pot needs to be reasonably low impedance (i.e. 5–10kΩ) to override the normal circuit conditions.

Many applications require more power. To power-up the oscillator and drive sections of a switch-mode power supply a supply of 9V–19V is required; to drive VCR motors under mechanical load calls for several hundred milliamps; and to operate a complete VCR or camcorder a low-impedance and closely stabilized supply capable of sourcing up to 3 or 4 amps is needed. It can also be used for charging batteries when required. The ideal instrument for all these requirements is a mains-powered PSU with built-in voltmeter and ammeter – the latter is particularly helpful in diagnosis. A maximum d.c. output voltage of 15V or 20V is required, with continuously variable control in the form of coarse and fine rotary pots or step-switched output voltage with vernier control.

To prevent accidents and damage to circuits, components and motors, some form of pre-set current limiting is useful. When fitted, and with the output voltage control set to maximum, a constant-current source is available. In the absence of any form of variable current limiting make sure that the PSU is protected against accidental short-circuit, overload and 'back e.m.f.'.

Frequency counter-timer

The counter/timer, like the DVM, has a digital readout, but much greater scale length – typically 7½ digits. Some are incorporated with a DMM in a 'measurement centre', a dual-purpose instrument. The frequency counter gives a readout of the frequency or period of the applied waveform. A general-purpose instrument will count up to 100MHz, more with a pre-scaler.

The accuracy of the instrument depends on two

main factors: the stability of its internal reference crystal, and the gate time selected by the user. Typically three gate times are offered: 0.1s, 1s and 10s, giving progressively greater accuracy, but longer settle-down time. The internal oscillator stability and ageing characteristics determine the price of the instrument to some degree, but for workshop (as opposed to laboratory) use, very high expense and accuracy cannot be justified. Inexpensive instruments generally have a semi-accessible trimmer to adjust internal clock frequency, permitting checking and re-calibration whenever required by reference to the field (50Hz), line (15.625kHz) and colour subcarrier (4.43361875MHz) frequencies within a TV set locked to a broadcast transmission.

Counters are mainly used for checking and setting up the crystal oscillators used for down- and up-conversion of the chroma signals in VCRs. Other applications include shaft-speed and tape-speed setting; adjustment of clock oscillators in VCR displays and camera/camcorder sync/subcarrier generators; checking the operation of digital counters and other logic circuitry, etc. It is important to realize the capabilities and limitations of the instrument, especially where the waveform being checked has more than one component, or is subject to frequency changes during the sampling period.

Test pattern generators

In recent years the need for specialized test patterns has declined as picture tubes have become self-converging, and as colour decoder adjustments have shrunk almost to vanishing point. The main requirements now are for accurate edge-castellations to set picture centring (i.e. line-phasing control), a central circle for adjustment of scan amplitudes and linearity; and a step-wedge in black-and-white for setting grey-scale (in most modern sets black-level adjustments for the three guns is carried out automatically within the decoder chip, leaving just the drive controls to be set manually for neutral highlights). A plain white or red pattern is required for check and adjustment of beam landing (purity). Useful for tests of focus and video bandwidth is a *multiburst*, which was described, along with several other pattern configurations, in the test-cards section of Chapter 1.

A fundamental pattern, required from any generator used for service work, is the standard colour-bar. The oscillograms published by equipment manufacturers for use in servicing and setting-up signal stages are based on the standard colour bars, though it's important that the standards used for reference are the same – or that allowance is made for any difference. The three forms of colour bars are illustrated in Figure 2.3.

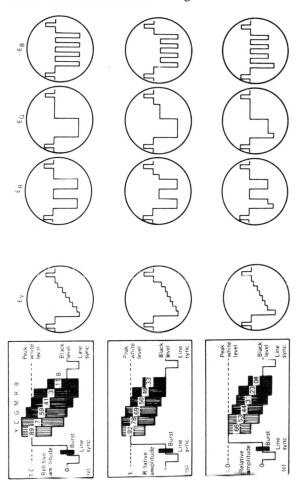

Figure 2.3 *Standard colour-bar waveforms: (a) 'Basic' signal 100% amplitude, 100% saturation, as generated by most commercial pattern generators; (b) 100% amplitude, 95% saturation as used by the BBC; (c) 75% amplitude, 100% saturation, used by the IBA and the EBU*

A wide variety of pattern generators is available, ranging from pocket-size 'spot-frequency' generators of setting-up patterns to elaborate mains-powered instruments capable of generating a complete composite test-card similar to those illustrated in Figures 1.1 and 1.2. Other types, again intended

15

primarily for workshop use, are mains-powered 'universal signal sources' offering a wide variety of test-patterns, including some of the 'Anti-PAL' types described in connection with the PM5544 test pattern in the previous chapter. Amongst the patterns will be plain fields of primary colours and 'split field' colour bars with a horizontal band of white to act as a reference when examining TV and VCR signal waveforms at field rate – a useful feature. Comprehensive pattern generators, too, generally have a choice of output channels and (in some instruments) bands, with calibrated output levels and switched r.f. attenuators. Sound modulation is also incorporated, and the most useful and versatile generators offer different audio f.m. carrier frequencies to suit different countries and standards. Some also have provision for TTL-RGB output for testing monitors.

Logic probe and pulser

These devices are basically computer test instruments, and are relevant to the control, syscon and logic sections of TVs, VCRs, disc players and their peripherals. These slim hand-held instruments are designed to inject (pulser) and extract/indicate (probe) digital logic states, including fast pulses. They can be used for testing gates, flip-flops, counters and other digital circuitry. The state indications are given by body-mounted LEDs, and in some probes by a variable-tone audio output. The devices are generally powered from the equipment under test via clip-leads, and can work with any types of logic, including TTL, MOS and CMOS.

The pulser should be capable of generating either a single pulse (typically 5μs wide) or a continuous pulse train at about 5Hz rate, which is easily detectable by eye in an indication LED. It should be able to operate on supply voltages between 5V and 15V with protection against incorrect polarity, and have a warning LED to indicate excessive output current – the output impedance at the probe has to be low enough to override existing conditions at the node to which it is connected.

The probe gives an indication of conditions at the point it is touched to in the form of (typically) a red LED for high (logic 1); a green LED for low (logic 0); and perhaps an orange LED to indicate the presence of pulse activity. Used alone, or in conjunction with the pulser, the logic probe saves a

lot of time when tracing faults in logic circuitry. A good probe should be capable of operating at frequencies above 10MHz, and detecting trains of pulses with widths down to 50ns; the pulse duty-cycle is indicated by the relative brightness of red and green LEDs. Figure 2.4 shows a pulser and probe in use in the syscon section of a VCR.

Figure 2.4   *Hand-held digital pulser and probe in operation*

Signal strength meter

Used for checking the level of broadcast signal available, strength meters are these days more the province of the aerial rigger than the TV/video engineer, especially since the number of terrestrial broadcast stations now exceeds 850 in the UK, and good strong signals are available at most sites. Modern TV and VCR tuners can produce acceptable pictures from signal levels as low as 200μV, but 1mV is the level which should be aimed at for noise-free reproduction of pictures.

Signal-strength indicators come in two forms. Simplest is the traditional meter, an instrument that gives a readout, in μV and mV, of the signal level from each station tuned. Its most common use is to 'peak-up' the carrier strength as the aerial is swung to and fro before being clamped.

The alternative – and much more sophisticated and expensive – form of signal strength meter is the *panoramic monitor*, a form of spectrum analyser. Here a varicap tuner is made to continuously scan the bands of interest with its output plotted on a scale of channels along the band. By this means, many new factors are brought to notice: the individual strengths of sound and vision carrier for each channel, the relative amplitudes of each of the channels/broadcasts of interest, and the presence of any interfering signals or spurious carriers. By adjustment of sweep position and amplitude ('shift' and 'zoom' functions) individual carrier pairs or groups can be selected and enlarged for close study.

Video-head and LOPT testers

Both VCR heads and LOPTs are inductive components; a check of their inductance is a good indication of their condition. Video heads are subject to wear, and line output transformers to the development of S/C turns as a result of overheating or internal discharge. Both are expensive components for which the only sure test is substitution – difficult if the replacement part is not in stock! The testers use an a.c. pulse train to check inductance: in general the results are presented in the form of green (good) and red (bad) LED indicators on a LOPT tester, and as an analogue meter readout (poor/borderline/good) in a VCR head tester. These instruments are not infallible, but used according to the instructions (and with experience of them) give a good and usually reliable indication. As with some of the instruments to be described shortly, their usage depends entirely on the habits and attitude of the engineer.

Isolation transformer

Many TV sets have a live chassis, that is one whose metalwork and internal ground line take up some potential with respect to true earth when running from the mains. Even where mains isolation is provided by a chopper transformer within the set it is often necessary to work on the PSU section. In all these cases, for safety, the set needs to be run from an isolation transformer. For modern TVs a 250VA type is adequate, but to cater for all sets that may be encountered a 500VA type is recommended. For bench use the transformer can be permanently installed and wired to a specially marked 13A socket; for field use, portable isolating transformers are available.

Variac

A variac is a mains auto-transformer with continuously variable turns ratio up to about 1:1.1. It has many uses in providing an a.c. supply at any

voltage from zero to full mains potential. Its main use, however, is in providing a 'soft' and safe start for power supply units under test and investigation, as the next chapter will show. In any equipment where there is danger of component destruction, overheating or overload at switch-on (i.e. direct-coupled output stages in high-power audio amplifiers) the variac can be used to gradually run-up the input voltage while monitoring voltage and current conditions.

Variacs come in various current ranges; the higher the rating of that used (up to 10A) the 'stiffer' and better-regulated the output. Ideally a voltmeter and true-r.m.s. ammeter should be incorporated.

Vectorscope

This instrument has a circular screen and is basically a measurer of phase. It is a somewhat specialized instrument mainly used for setting up the encoders of TV cameras and camcorders, especially those with analogue (i.e. tube-type) optical pick-up devices. The circumference of the screen represents one complete cycle of carrier (in practice, usually colour subcarrier at 4.43MHz), and with no input signal there is a single spot of light at screen centre. The amount of spot deflection away from screen centre indicates the amplitude of carrier (i.e. saturation) and the angle at which it is deflected indicates phase relative to subcarrier reference – hence hue. When the camera is scanning a standard colour-bar image the vector-scope's light spot is deflected in turn to well-defined spots on the display. The phase and amplitude of the chroma signal's *burst* component is also shown.

Signal injector

The signal injector, seldom used now, is an analogue equivalent of the logic pulser described earlier. It contains a squarewave oscillator powered by an internal battery. The harmonic-rich output can invoke a response in circuits working at audio, video, i.f. and even r.f. frequencies. The idea is to isolate a faulty stage by injecting a signal progressively 'backwards' from the loudspeaker or picture tube. It is not really appropriate to modern equipment and circuits; use of a function-generator or signal-generator, or test-pattern injection, is more scientific.

Reference light source

The colour of white light is difficult to define by eye, and it is important that TV displays are set up to give a white colour-temperature corresponding to illuminant D. To assist with this, some engineers and workshops feel it worthwhile to buy a reference light source. It consists of a mounted fluorescent tube with a closely specified fluorescent phosphor colour, surrounded by progressively denser neutral-filters to give the correct reference light at three or four different levels. The device is held close to the TV screen while adjusting R, G and B drive and black-level presets to achieve a colour-temperature match.

Degauss wand

The manual degausser is merely a coil of wire powered from the a.c. mains supply. It generates an alternating magnetic field which is held close to magnetically critical metalwork (i.e. picture-tube shadowmask, shield and nearby steel objects) and then withdrawn to a distance. The decaying magnetic field demagnetizes the object treated, and should ensure correct beam landing to give a properly *pure* picture, colour-wise.

Test bridge

Passive components like resistors, capacitors and inductors can be checked with great accuracy by means of a test bridge. It may have a meter or digital readout with auto-balancing (modern design); or have a 'null' indicator to enable the component value to be read from the rotary dial with which manual balance is achieved (traditional type). Some engineers make much use of these, while others find no use at all for them.

Insulation tester

For many years insulation and dielectric strength was tested by a *Megger*, incorporating a hand-cranked high-voltage generator. Safety checks for compliance with BS415 and BEAB standards demand certain applied voltages for specified periods. These can be applied and leakage measured by electronic testers in which voltages of 500V, 1kV, etc. are generated by electronic means. On an analogue or digital readout resistance of 100MΩ and more can be measured.

IR radiation testers

The two main sources of infra-red light which concern the engineer are those from the tape-sensor LED in a VCR; and from a remote control handset. In both cases the light is invisible to the eye. For tape-sensor LEDs a phototransistor probe can be made up and mounted on a 'wand' to intercept and measure IR emission; it's usually simpler to use the end sensor LEDs themselves, however, by alternately exposing and shielding them from the central emitter while monitoring their change in resistance.

For checking the radiation from an IR handset a detector/amplifier can be taken from a scrap TV or VCR, and its output applied to an oscilloscope. A useful 'quantative' check is given by a device called a *magic mirror* infra-red detector. This visiting-card-size device reflects IR radiation in visible form, so that an orange glow is seen in subdued light when a working handset is pointed at the active surface. The strength of emission can be judged with experience. A photo of one of these devices is the subject of Figure 2.5.

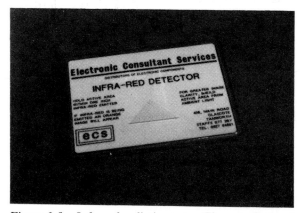

Figure 2.5 *Infra-red radiation tester. The central triangle reflects IR light in visible form*

3

POWER SUPPLIES

In home-base TVs and VCRs the basic energy supply is domestic a.c. mains, and it is the job of the PSU (power supply unit) to convert this to a closely stabilized d.c. voltage to operate the various circuit sections. Especially in TVs the PSU is one of the most troublesome and fault-prone sections – indeed this and the line timebase section between them account for most of the faults encountered so they will be treated at length in this and the next chapter.

Power-supply circuits come in a large variety of types and basic configurations, which enjoy various degrees of reliability and ease of servicing! For mains use there are two basic types of power supply: those using a 50Hz mains step-down transformer followed by a rectifier and series stabilizer section, much used in VCRs, monochrome and some small-screen colour TVs, and a few designs of larger-screen CTVs and monitors; and those working on a *switch-mode* principle, by far the most common in TV sets, and finding a more limited use in VCRs. Of the switching-type PSUs, again two broad classes can be distinguished: those switching at mains (50 or 100Hz) rate, generally using thyristors in now-obsolescent designs; and those switching at a high rate (i.e. 15.6kHz line-scan frequency) which are more efficient and account for virtually all current and recent designs

of CTV. All these types of PSU are outlined in Figure 3.1. Each will be discussed in turn, to an extent depending on how reliable they are, and how frequently they are encountered in service work.

Overview of SMPSUs

Switch-mode power supply units have four essential sections, outlined in the block diagram of Figure 3.2. First (a) comes a mains rectifier and reservoir capacitor to produce a high (typically 320V) primary operating voltage. Next (b) comes a switch, generally in the form of a transistor, opening and closing at regular short intervals to regulate the energy fed to the third section (c): a reservoir of energy which may be inductive, capacitive or a combination of both. The fourth component (d) is the control section, which varies the duty cycle (mark/space ratio) of the chopper switch according to feedback information from the secondary reservoir (output line) across which the load is connected. The regulator/control section also incorporates various 'safety' artifices: over-voltage and overcurrent protection, invoked in the event of malfunction in the PSU and the load respectively; and often a *soft-start* circuit to ensure that the PSU gradually runs up to full output voltage at switch-on. In the

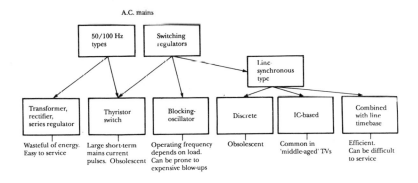

Figure 3.1 *Forms of mains-operated power-supply systems*

representative circuits of PSUs to be used as models for fault-finding we shall identify the sections corresponding to a, b, c and d in Figure 3.2. As we examine each circuit in turn, the various techniques of diagnosis discussed will not necessarily belong only to that circuit. Most of them have application to most PSU circuits, depending on the nature of the fault and the specific design of the PSU.

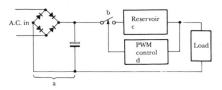

Figure 3.2 *Basic operation of switch-mode power supply*

Blocking-oscillator SMPSUs

A common configuration of PSU is the one given in Figure 3.3. Here the chopper control is carried out by the purpose-designed chip ICI, type TDA4600, and the frequency of operation of the circuit depends entirely on the loading: with no load at all Trl and the chopper transformer Tl oscillate at about 80kHz; with an excessive load (i.e. short-

circuit secondary diode D5) the frequency of operation drops to 1.5kHz, the mark/space ratio widens out to about 1:250 and the circuit will operate indefinitely under these circumstances. Normal operation finds the circuit running at about 25kHz. Basically the configuration is that of a blocking oscillator, in that base current is not fed to Trl until all the energy from the previous pulse is drained from the chopper transformer, as indicated by conditions on IC pin 2.

The stabilizing control loop is based on winding 3–5 of T1 whose pulses are rectified by D2 to produce a negative sample voltage across C2; this is applied to the sensing input of the IC (pin 3) via the set HT pot VR1. Insufficient negative voltage at C2 (due to C2 itself or D2) will increase the output voltage of the PSU and push VR1 'off the clock', but beyond a certain point (and before output voltages rise dangerously high) the TDA4600 goes into 'short-circuit (S/C) mode' and virtually shuts down the PSU. This is the same effect as when one of the transformer's secondary windings 6/7/8 is shorted: the sample voltage at C2 disappears. An oscilloscope, hooked to base or collector of Trl will, on close and careful examination, show very narrow (3µs) needle pulses at intervals of about 0.7 ms. This is a sure sign that the IC has shut down, and the short-circuit (S/C) condition may be easily

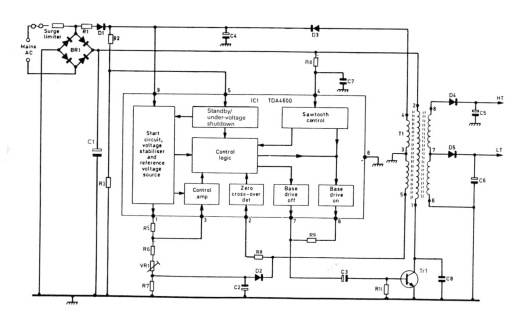

Figure 3.3 *Late configuration of blocking-oscillator power unit using an IC type TDA4600*

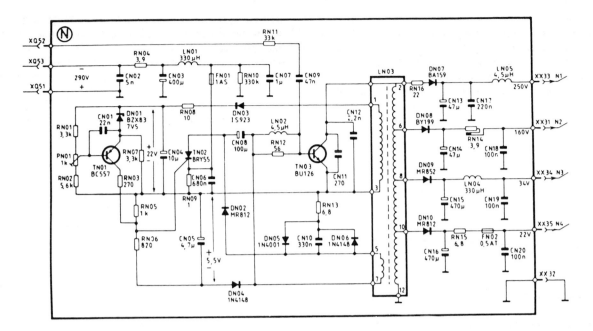

Figure 3.4 *Blocking-oscillator PSU using discrete components*

found: start by checking with an ohmmeter across C5 an C6 to detect a S/C line output transistor/ flyback diode or rectifier D4/D5. Other possibilities are a faulty D2 or C2.

The start-up supply for the IC comes from the mains via R1 and D1 to pin 9; complete failure of the set should lead to a check of these two components and the voltage across C4. Once established, IC operating voltage comes from T1 pin 4 via D3 and smoother C4; if these are faulty D1 cannot isolate R1 and the latter will overheat.

If the set is completely dead and the mains fuse F1 is blown, first check with an ohmmeter for S/C or leakage in the diodes of full-wave rectifier BR1, and chopper transistor Tr1. If the latter has failed, the problem could well be due to 'round-shouldered' drive (failure of R9 and/or C3) or to failure of R4, a high-value (i.e. 270kΩ) resistor on which is based the chopper drive timing *and* an over-dissipation-protection circuit for Tr1. The unfortunate thing is that failure of Tr1 can often 'take-out' IC1 or vice-versa, so it's wise to replace both (and check the other components mentioned above) in the case of Tr1 breaking down.

Pin 5 of the TDA4600, once the start-up phase is over, will shut down the PSU if its potential falls below about 2V. It has two purposes: mains under-voltage protection, and remote-control of standby operation. Failure of the PSU to perk up, then, may be due to IC pin 5 being held too low by problems in R2 or a faulty (stuck low) 'standby' control line.

Discrete blocking-oscillator

An older-fashioned (but still likely to be encountered) blocking-oscillator PSU is illustrated in Figure 3.4. The concept is due to Siemens, and has been used by many manufacturers (Decca, GEC, Grundig, Hitachi, Rank, Tandberg, etc.) over the years. Figure 3.4 (Luxor) represents several similar circuits of that period. This is a simple and reliable type, but can be difficult to service; it is an 'all or nothing' affair in which the transistor TN03, transformer LN03 and associated feedback and control circuits form an oscillator. If any part of the circuit fails the oscillator ceases to function, making diagnosis difficult. Failure of the feedback/regulating mechanism (based on LN03 winding 5–7) can cause the circuit to run out of control and destroy the chopper transistor or produce output over-voltages before blowing the mains input fuse.

In theory this type of circuit has an inherent form of overcurrent protection: a heavy load on any section of LN03 causes the oscillator to close down to a very low operating frequency at which it can safely run in the overload condition. The characteristic purring from the transformer in this condition should lead to a check of the secondary rectifiers DN07/8/9/10 for S/C, and particularly for a S/C or leakage on the 160V line, i.e. a faulty line output transistor or recovery diode.

Sometimes the protection mechanism does not work, and failure of one of the above-mentioned semiconductors can 'blow-up' the PSU in the same way as an internal fault. In this case the mains fuse will be found to be blackened and often shattered and the chopper transistor (usually a BU126 or BU326) short-circuit. If the fuse and transistor are replaced and the set switched on there is a good chance that instant destruction of the new components will take place. To prevent this it is wise to check all the other semiconductors (TN01, TN02, diodes, etc.) with an ohmmeter before switching on, and to start up from a low mains voltage using a variac. In the event of a high current flow, further checks are necessary. When using a variac, bear in mind that this type of circuit will not start up with a very gradual 'tail-on' of mains voltage – a kick-pulse via CN09 starts up the circuit illustrated here; other versions use a start-up transistor switch. If the circuit is working correctly, kick-off is generally possible with a mains input voltage as low as 160V or 180V. An alternative is to wire an ordinary 60W mains light bulb in series with the a.c. mains input or in place of fuse FN01 to limit the current for test purposes. With a low applied input voltage a working PSU and set should produce an output, albeit low, to prove operation of the PSU. This may not result in any light on the screen (or a small dim raster may take a long time to appear) because the picture-tube heater will be under-run.

In this type of power supply, intermittent failure to operate – or a tendency to 'blow-up' at irregular intervals – may be due to dry-joints in the PSU generally, and particularly at the pins of chopper transformer LN03. Other points worth checking for this fault are the pull-down capacitor CN08, the 1Ω sampling resistor RN09 and regulating thyristor TN02. Incorrect output voltage or adjustment range of PN01 may be caused by problems in the regulating circuit: reference 7.5V zener diode DN01, sample-source components DN03/CN04 or comparator transistor TN01.

Some versions of this circuit have a 'crowbar' trip to sense overvoltage: it consists of a sensing zener diode and thyristor which shorts out the primary (320V) or secondary supply. Blown mains fuses, then, should direct attention to this as well as the bridge rectifiers. Some versions, too, do not like working with no load; a 60W 240V light bulb is an adequate load in place of the line output stage.

IC-driven SMPSUs

This numerous class uses an IC to provide a square-wave drive to the chopper transistor via a driver stage and step-down transformer. Control is effected by varying the mark/space (M/S) ratio of the chopper drive waveform. These types of PSU are easier to service than blocking-oscillator types, and are less destructive when they do fail, stemming from the fact that their sections are less interdependent.

One form of IC-driven SMPS is illustrated in Figure 3.5. Here IC 801 provides a drive at its pin 6 for application to chopper transistor T12 via T11 and transformer L8/9. IC801 is powered from the primary d.c. supply via R79 and C40, and T12 likewise via R82. For diagnostic purposes this IC can be operated by itself, with the chopper transistor drive (or emitter feed) disconnected. So long as its 12V supply at pin 1 is maintained by internal or external means it should be possible to trace with an oscilloscope the chopper drive waveform through to T12 base. In all testing on this type of power supply the − 320V line is the effective 'ground' point, and its high potential with respect to the TV chassis (and to true earth without a mains isolating transformer in use) should be always borne in mind for safety's sake.

Complete loss of action in this PSU is tackled by first checking the mains fuse. If it's open-circuit (O/C) check the bridge rectifier diodes, degauss components and mains filter capacitor – also if necessary the chopper transistor T12. Failure of the latter is more likely to blow F3, and if T12 has shorted c–b or c–e a likely cause is failure of pull-up resistor R80 in the base feed of the driver transistor. This chopper circuit is inherently safe for the rest of the set due to the fact that L11 is grounded; serious PSU failures cause the main secondary supply at C51 to sink to ground rather than rise.

As well as a soft-start feature, IC801 has built-in overload protection artifices, and the most common fault symptom in this type of PSU is a 'pumping' effect – there will be five or six pump cycles before the set shuts down altogether, indicated by the

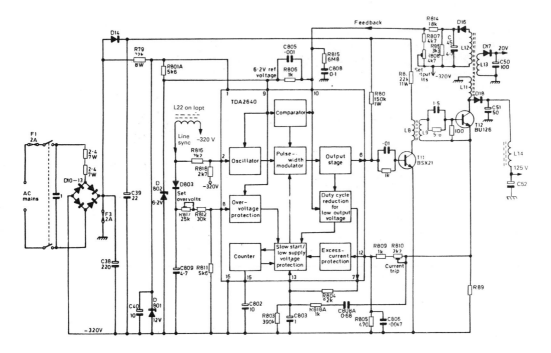

Figure 3.5   *IC-base line-synchronous chopper power-supply (ITT)*

presence of about 6V on IC pin 15. The two instigators of pumping symptoms are overcurrent and overvoltage. Dealing first with overvoltage tripping, the giveaway is the appearance of EHT and a rushing sound from the speaker on each pump cycle. The check procedure for this is to hook a d.c.-coupled oscilloscope across C52 (or its equivalent in other circuits of this type) and check the peak output voltage on each 'plop'. If it exceeds normal voltage (here 125V) the IC's reference (D802) and feedback (from C45 via R814 etc) voltages should be checked, and R808 re-set.

Much more common is overcurrent tripping, with little sign of life throughout the rest of the set. The usual cause is overload on the + 125V line. Disconnecting L14 proves this by restoring life to the PSU, with a light audible whistle and the + 20V line present at C50. In this case the line output stage should be investigated, initially with an ohmmeter – see the next chapter. For a realistic 'dummy load' a 60W 240V light bulb can be substituted for the normal load on the + 125V line – it should light at about quarter brightness. If overcurrent tripping continues with no excessive load, current sampling resistor R89 and the setting of overcurrent pot R810 are suspect.

PSUs of this type and vintage are designed to run synchronously with the line output stage to avoid 'beat' interference effects. If striations or moving patterns are present on screen check the synchronizing system: L22/R816 etc. in Figure 3.5. In some types of PSU lack of sync between PSU and LTB can give rise to an audible squeal.

Intermittent shutdown or failure to start may typically be caused by the IC itself or failure of its supply (scope shows no drive); or an O/C secondary rectifier D18 (audible whistle and 20V line present).

Mains-isolated type

A later design of PSU with IC chopper drive is shown in Figure 3.6. This is very similar in principle to the one just described, though four transformers are used for mains isolation. T702 operates at 50Hz to provide 'start up' supplies for the chopper control chip, remote control receiver and the driver transistor, replaced for the latter, once the PSU is under way, by the + 22V line derived from T705 secondary. T703 is a ripple (hence current) sensing transformer, T704 the chopper drive coupler and T705 the chopper

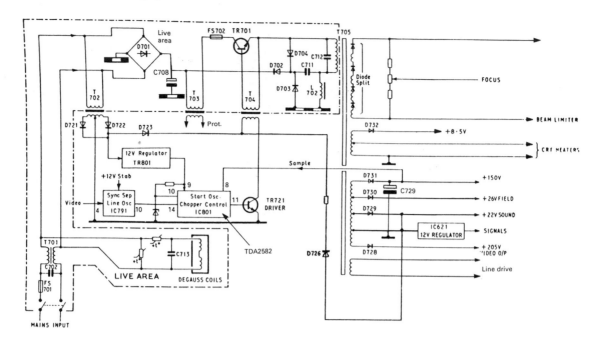

Figure 3.6 *Simplified diagram of mains-isolated PSU (Ferguson)*

output transformer. All the circuitry outside the dotted box is isolated from the mains and may safely be connected to true earth.

Fault-finding here largely follows the principles already described. A blown mains fuse would probably indicate failure (S/C or leakage) in the mains bridge rectifier D701, filter capacitor C702 or (more rarely) a problem in the degauss network, bifilar choke T701 or reservoir capacitor C708. If fuse 702 is blown, the likelihood is that chopper transistor Tr701 is S/C, and a search for contributory causes (ie D702/3/4, C712 etc) is appropriate before replacement of TR701.

Most other faults take the form of PSU tripping. This may arise from problems on the primary or secondary sides of the chopper transformer T705. If there is virtually zero voltage rise across C729 (150V reservoir capacitor) the problem is *within* the PSU so long as D731 is intact. PSU-internal faults may be D702/3/4 S/C or O/C; L702 O/C; C711/2 faulty; or PCB faults like dry-joints or hairline cracks.

Fault diagnosis can be difficult if the PSU is tripping very rapidly. In those circumstances, ground pin 8 of IC801 to maintain a low chopper

duty-cycle and hence limit the peak chopper collector current to a safe value. Unless there is a direct fault on the primary side of T705, the PSU should now function (reduced output levels) without tripping.

The majority of tripping faults stem from causes outside the PSU, revealed by an attempt to rise – on each 'plop' – of the secondary voltages: the 150V line may typically achieve 10V–40V, depending on the severity of the load. Here the manufacturers, Ferguson, recommend unloading circuit sections one at a time until the tripping stops, in this order: EHT and focus unit, to eliminate faulty focus unit or excessive tube anode current for any reason; plug 14 (disconnects line output stage) to check line output transistor, transformer and associated components; plug 15 (takes out field scan coil) to check field timebase and output transistors; plug 10 (checks for shorts in RGB output stages); and plug 6 (checks 12V stabilizer and line-generator chips). This progressive load-shedding system is a standard approach to tracing an overload problem, whether the afflicted stage is a PSU or LTB section. The plug reference numbers here refer specifically to the Ferguson TX10 chassis.

'Discrete' SMPSU

A popular configuration for chopper power supplies was the discrete circuit, operating to the same broad principles as those described above, but using transistor circuits to drive the chopper transistor. Figure 3.7 shows an example. The primary energy supply comes from bridge CR901–4 and reservoir C909. The parallel chopper is TR906 which switches the bottom end of T902 primary to ground. A 110V output is developed by CR908 across C925 to form the stabilized output.

This circuit is a synchronous one, driven by line output pulses, entering at the left. They are integrated by R910 and C915 to produce a sawtooth waveform at TR903 base, an amplified and inverted version of which appears at TR903 collector and at TR904 base. The point in the sawtooth at which TR904 is turned on is governed by the standing bias on its base, and this comes from TR907 under the influence of the 110V line voltage. As the output voltage drops, TR904's threshold point

moves to increase (via driver TR905) the duty-cycle of chopper TR906, restoring the output voltage to normal. The thick-film assembly CP901 provides both a zener reference voltage for TR907 emitter and a precision voltage divider for its base, removing altogether the need for a pre-set pot to set output voltage level.

The parts at the bottom centre are for safety protection. All is based on the action of the two transistors TR908/9 which are wired in thyristor configuration and will, when TR908 is pulsed on, latch up in conduction and ground R941 to stop TR904 and shut down the PSU altogether. Conditions are not sampled again until the set is switched off and on once more, when shutdown will again take place if the overload condition is still present, a point worth bearing in mind when investigating a 'dead set' symptom: a single burst of energy will be present at switch on with this type of protection. Sampling points for safety cut-out are beam current via CR712, line pulses (hence EHT voltage) via CR709, and line output transistor emitter current via R722 and zener CR713. With

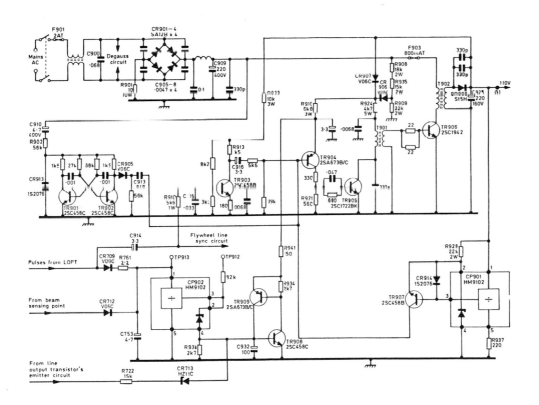

Figure 3.7 *Chopper PSU using discrete components and thick-film reference modules (Hitachi)*

bulb or variac precautions the cut-out system can be disabled by disconnecting R941; this permits checking of the overload conditions.

Because the circuit depends on line drive pulses for its operation, and the line timebase depends on the PSU for power, some means of starting up the PSU is required. A kick-start mechanism is provided by the astable multivibrator TR901 and TR902. Their oscillation provides pulses into TR904 base so long as C910 is charging from the + 320V line. Here, then, is another possible cause for the same symptom as the safety cut-out described above: a burst of energy at the instant of switch-on only. This will be due to the line timebase not working, and unable to take over from the start-up system. On the other hand failure of the kick-start mechanism will result in the 'dead-set' symptom, save only for the operation of the degauss system, which can sometimes be heard as an audible buzz. If it appears that the kick-start astable is not running, its operation can be sustained by shorting C910 and checking at CR905 with an oscilloscope. If this restores normal operation which continues when the shorting link is removed, the chances are that C910 itself is O/C. If not, the square-wave pulse drive can be traced through TR904/5/6.

Kick-start systems, usually relying on the charging of a capacitor, were often used to start PSUs, line oscillators, and occasionally line and chopper drivers. It is often necessary to override the kick system, and this can be done by bypassing the capacitor by a 5W or 10W resistor of about 5.6kΩ, or by applying a suitable voltage (typically 12V or 18V) from an external supply. This maintains oscillation while checking and fault-finding.

Series chopper

The types of chopper so far examined have been *parallel* types, in which the transistor switch gates a 320-odd volt supply to the primary of a transformer. An alternative system, used in the first-ever SMPSU in a domestic TV (Ferguson 3000 series) is the *series chopper*, a configuration which lingered on for many years with various manufacturers.

A skeleton diagram of a series chopper system is given in Figure 3.8. At each 'on' pulse of TR1, energy is driven into storage choke L1 and the load. When TR1 switches off, D1 clamps the left-hand end of L1 to ground and the choke's stored energy flows out into the load. C2 acts as reservoir/smoother for the secondary supply: the regulator/

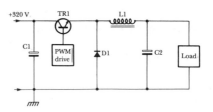

Figure 3.8 *Working principle of series-chopper circuit*

driver circuits operate in the same way as those we've already examined to produce a variable duty-cycle switching drive. In practice the disadvantages of this type of circuit are: (a) the full 320V primary supply could be applied to the load in the event of the chopper transistor shorting internally or failing to turn off; and (b) only a single output line is available at a set voltage, typically 100–160V.

These problems are overcome, respectively, by fitting a crowbar or other form of excess voltage sensor (ZD1 and TH1 in Figure 3.9) to the output line; and deriving secondary supplies from the line output transformer – see Chapter 4. The use of a driver transformer in some designs (i.e. Figure 3.9) obviates the risk of the chopper getting permanently biased on, and the derivation of line drive from a secondary winding on the chopper choke gives automatic shutdown – to safety – of the set if the chopper transistor should go S/C. In the design of Figure 3.9 the chopper drive is synchronized with TV line rate. Even so, series chopper circuits have now fallen into disuse for the reasons given in the last paragraph, and the fact that they are 'live-chassis' systems.

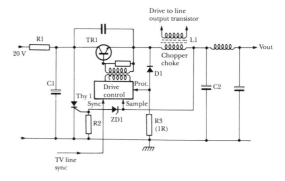

Figure 3.9 *Series chopper with line sync and protection system*

In this type of circuit, a shorted chopper transistor TR1 or clamp diode D1 will blow the mains input fuse violently. Where a crowbar circuit is fitted, a violently blown mains fuse is often the result of the crowbar itself firing, for reasons legitimate or otherwise. When nothing can be found wrong with the main circuit, false firing of the crowbar thyristor is usually due to internal leakage – in the thyristor itself or one of its gate-feed components, typically ZD1 in Figure 3.9. These remarks about false crowbar firing are relevant to any type of PSU circuit you may find them fitted to.

Mains-rate thyristor regulators

Again this is a type of circuit no longer used, but is included here because many sets using it may still find their way onto the repair bench. Half-wave or (Figure 3.10) full-wave circuits may be encountered; both types draw short, heavy gulps of current from the mains, and so require a large (500W or more) isolating transformer to power them on the test bench. The thyristor control element(s) THY1 is in series between the mains supply and its load, though blow-ups due to S/C control thyristors are quite rare.

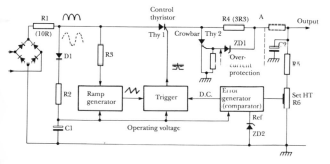

Figure 3.10 *Principle of thyristor-controlled power regulator*

The basic operation depends on the fact that the output voltage (energy fed into large reservoir capacitor C2) depends on the point in the mains cycle at which regulator thyristor THY1 is turned on; it turns off at each zero crossing of the mains waveform. Thus the timing of the trigger pulse to its gate determines the output voltage. The firing pulse is generated by a trigger circuit with two inputs: a mains-rate ramp and an error voltage, the latter derived by measuring the output voltage at point A against a reference potential at zener diode ZD2, generally rated at 6.2, 6.8 or 7.5V. The trigger section produces an output pulse whenever its two inputs coincide in voltage. Since the ramp

has a constant characteristic this pulse advances and retreats in phase according to the measured output voltage at C2, thus regulating the output to a point set by control R6.

Very common in this type of PSU is a blown mains fuse, and its replacement will probably restore normal operation but then itself blow prematurely. The crowbar thyristor THY2 is by far the most common instigator of this, and may be (a) triggering falsely for the reasons outlined above (voltage-sensing zener diodes rated above 100V – here ZD1 – are particularly suspect) or (b) legitimately firing because PSU output voltage sometimes wanders up. The latter may be due to leakage in the control thyristor THY1, change in value in the upper resistor (R5) in the 'set HT' potential divider chain, or thermal drift in the associated reference zener ZD2.

Some thyristor PSUs (i.e. Decca) have an over-current trip which works on the crowbar thyristor also. The sampling resistor is shown dotted at point A in Figure 3.10. Here a violently-blown mains fuse may also be due, then, to a faulty EHT tripler, line output transformer etc. A 60W light bulb is useful here. It can be put in series with the main HT feed to the line output stage to limit the current to a safe value for testing; and then if no fault is obvious it can be fitted in series with the crowbar thyristor itself as an indication of when the latter has fired. If when the lamp comes on the set continues to work normally false triggering is indicated. The lamp trick in this case robs the set of its normal protection, and caution is recommended here to avoid damage to PSU and load. Operation from a variac, or with a bulb-load in series with the mains feed to the TV makes the exercise safer. For this application, one 100W or two 60W bulbs in parallel are suitable.

A 'dead set' symptom with this type of power supply, where the mains fuse and bridge rectifier are intact, should lead to a check of the operating voltage for the triggering/control circuitry. Typically 20–40V, it's usually derived from the raw 320V line via a dropper resistor R2 and decoupler C1. These and any zener diode at the 'bottom end' (i.e. across C1) should be checked if the voltage is absent; a quick indication of current flow in the dropper resistor is to check that it's warm after a run. A puzzling no-go symptom arises when the control thyristor's cathode is not grounded to a.c. due either to an O/C in the reservoir capacitor C2 or some small resistor R4 between the two. Because the thyristor cannot fire without gate current, and because gate current depends on a path to earth, no voltage at all appears at the control thyristor cathode under these circumstances. The oscilloscope will show a voltage pulse at the thyristor's gate and an identical pulse at its cathode. Other no-go faults involving failure of the regulating thyristor to fire are generally due to lack of a triggering pulse and can be traced with an oscilloscope in the low-level control/regulation circuits, which in sets of this vintage invariably use discrete diode/transistor/ zener technology; with the scope triggered at mains rate the gate-control pulse should be seen to advance and retreat in time/phase as the set HT pot is manipulated. If the pulse is missing the ramp has probably disappeared: check ramp-forming resistor R3 and associated components. Some circuits use a zener to clip the mains waveform in the ramp generation circuit – check that it is saturated where relevant.

SMPSU fault-finding chart

Each of the PSU designs we have examined so far is intended to represent one *class* of circuit, from which it should be possible to identify and tackle the type of PSU under repair. Regardless of the type of PSU in use, there are several 'ground rules' in fault diagnosis, and these are set out in generalized form in Figure 3.11. Summarizing the chart, it's first necessary to establish that primary operating voltage is available to an undamaged chopper

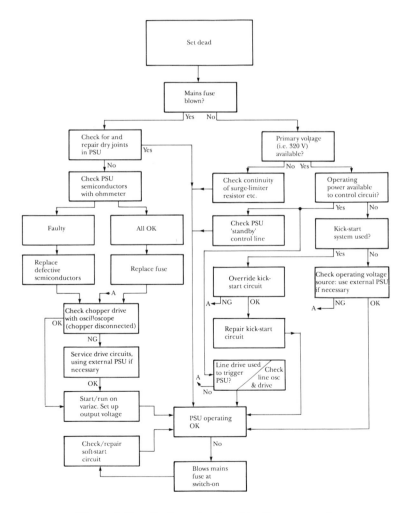

Figure 3.11   *Fault-tracing in a 'dead' power supply*

element. Next (where applicable) comes the need for operating voltage to the drive/control circuits, even where they only consist of a kick-start feed to a blocking oscillator. When this is present, no results suggest lack of chopper drive, investigation of which requires a multimeter and an oscilloscope.

Pumping symptoms are usually the result of a healthy PSU responding to an external fault. Ohmmeter checks and the disconnection of circuit sections in turn (starting with the line output stage) will generally reveal the faulty section. Some approaches to the 'pumping' symptom are given in Figure 3.12.

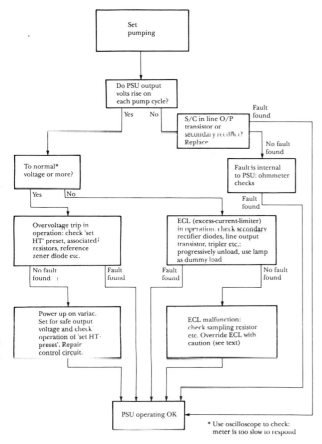

Figure 3.12 *Dealing with 'pumping' and 'tripping' in a PSU*

Use of lamp bulbs

In the foregoing text, several references have been made to the use of ordinary household BC-type light bulbs. They make inexpensive high-wattage resistors, with the advantage of giving a strong visible indication of the current flowing. Singly or in combination they are useful as series current limiters in PSU and LTB applications, and as a substitute load for the set while PSU testing. At a typical d.c. output line of 125V the bulb's power dissipation is about one-quarter of its rating. Thus a 60W bulb absorbs 15–20W in normal circumstances, a reasonably representative load. A 100W bulb is perhaps more representative of the demand of a large or wide-angle screen TV.

Capacitor failures

The two most common failings of capacitors in PSUs are surge currents in mains filter capacitors (examples are C702 in Figure 3.6 and C900 in Figure 3.7); and drying up or ageing of electrolytic capacitors used for reservoir and smoothing applications. Mains filter capacitors range from 0.0047μF to 1μF and are fitted across the mains input, downstream of the anti surge fuse, to intercept spikes and interference travelling to and from TV and VCR equipment. Failure blows the mains fuse, and the smaller the fuse rating (especially in VCRs) the more likely the trouble. Some manufacturers recommend reducing the size of the capacitor and/or increasing the fuse rating. Some technicians do this whether or not the manufacturer recommends it! In VCRs, the author generally fits a 0.0047μF 250V a.c. capacitor of class X, and has had no repetition of trouble from either fuse-blowing or interference.

Loss of capacitance in electrolytics is often encountered in older sets. Where the capacitor is the primary reservoir (i.e. CN03 in Figure 3.4) the effect is to superimpose a large 100Hz ripple on the chopper supply – and, possibly, on the secondary output, to give a heavy hum from the loudspeaker and a double-waisted picture like that shown in Figure 3.13. Loss of capacitance in secondary smoother/reservoirs (i.e. CN14 in Figure 3.4, C45 and C51 in Figure 3.5) has more subtle effects, ranging from a ripple on the supply line – screen effect in Figure 3.14 – to low output voltage or pumping and operation of the overvolts trip system. This happens where voltage sampling is taken from across the low or O/C capacitor, and is upset by the pulses appearing there. These situations are usually easily seen with an oscilloscope hooked to the line in question: normally the peak-to-peak ripple

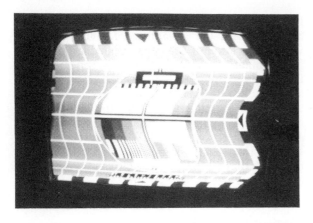

Figure 3.13 *100Hz ripple on picture due to hum-modulation of internal supply line*

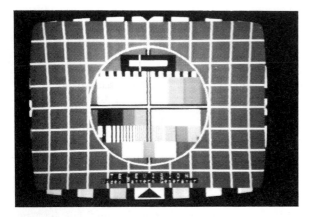

Figure 3.14 *'Serration' effect of high-frequency ripple on a TV's internal supply line. The jaggedness of picture verticals is less obvious here than the black tearing lines across the step-wedges in the central circle*

voltage does not exceed 2 or 3 per cent of line voltage.

Fuses

Most fuses used in PSUs are of the glass 20mm type, and except where there is an official manufacturer's modification it's important to replace like with like. Where a PSU problem involves constant fuse blowing during diagnosis it is cheaper and more convenient to wire a thermal cut-out (of the type used in older TVs) across a blown fuse, then re-set the button as necessary. A rating of 2A is appropriate

for TV work. Careful examination of a blown fuse reveals much about the cause of its failure. If the glass body is cracked, broken or fully blackened inside the likelihood is that a very large current ruptured it, usually via a mains filter capacitor, S/C rectifier or chopper transistor, or a triggered crowbar thyristor. If, alternately, the fuse is clean inside and the break in the element can be seen, it has failed 'softly' as a result of a relatively slight overcurrent or metal fatigue/ ageing in the fuse itself. In these circumstances replacement of the fuse generally restores normal operation, though a check of the fuseholder is worthwhile – any bad contact at the clips causes fuse overheating and premature failure.

Modern miniaturised equipment uses a small design of fast-acting fuse (ICP) about 5 × 3mm in size, and resembling a two-legged transistor. Of the two types, F and N, it's recommended to stock the 'N' type (both are expensive!) because this can be made to fit in place of either sort: bear in mind the requirements of BEAB, however, The rating numbers on these fuses must be multiplied by 40mA to give the rated current, so that ICP-N20 is an 800mA type, and ICP-F38 a 1.5A type. Occasionally these ICP fuses will blow for internal reasons, replacement permanently restoring correct operation. Even so, in VCRs (where they are most commonly encountered) a blown ICP may be the result of a faulty drive motor, or excessive current in a motor driving a too-stiff mechanism for cassette or tape loading.

Fusible resistors

In TV sets and to a lesser degree in VCRs, manufacturers fit fusible resistors to meet safety standards. They look like ordinary resistors and generally have values below 10Ω. They are identified on the circuit diagram by the △ or ⟨s⟩ symbol. Sometimes they fail for internal reasons, which may be betrayed by a pinhole on the surface or a crack in the body. Legitimate fusing of the resistor usually leaves a visible burn-mark, though even this can sometimes be caused by internal thermal runaway. A burnt resistor, fusible or not, is usually the result of excessive current, but a misleading situation can arise where a heavy *ripple* current flows due to failure (O/C) of a reservoir capacitor upstream. This a.c. current, not obvious in d.c. measurement, flows to a smoothing capacitor downstream of the resistor; this can happen in circuits working at mains, TV-line or PSU-operating frequencies.

HT jitter effects

An occasional problem with SMPSUs is output voltage jitter, giving rise to spasmodic changes in picture size. It can stem from problems in the line timebase or EHT-generator sections, but the PSU is more likely to be responsible, and can be proved so by monitoring its main output line voltage with an oscilloscope. Most possible causes of this effect are inside the stablization loop and should thus be nullified. Likely culprits, then, are the reference zener (DN01 in Figure 3.4, ZD2 in Figure 3.10 etc.) or the components in the set-voltage potential divider, where any electrolytic capacitor, thermistor or high-value resistor is suspect, as is the skeleton pre-set pot itself. Use of freezer aerosol and gentle heat from a hair-dryer will often find the culprit – see Chapter 20.

Replacement components in SMPSUs

In repairing mains-driven switch-mode PSUs it's important to replace faulty components with adequate types for maintenance of safety and reliability. Safety ($\triangle$ or $\langle \text{S} \rangle$) components have to be obtained from the manufacturers, and it's wise to get replacement semiconductors, capacitors etc. from the same source. This is particularly relevant to chopper transistors, where simple comparison of published data (i.e. as used in compiling the popular equivalents books) is not sufficient to ensure that the replacement will work correctly, run at normal temperature and not fail prematurely. Similarly the voltage rating of a replacement capacitor does not alone ensure that it is suitable: some capacitors have to withstand heavy ripple currents at high frequencies. In miniature equipments the physical size and shape of replacement components is also a major factor.

Camcorder mains adaptor/chargers

The most compact mains SMPSUs are those (provided with camcorders) made in the shape and size of the batteries they replace – and charge. High

efficiency is essential here to avoid heat dissipation, and the principles of operation are similar to those already described. Common cause of failure are O/C thermal fuses or opening of conventional fuses inside the unit. Before getting deeply involved in fault-diagnosis and servicing of these units it may be wise to check the manufacturer's price for a replacement unit.

50-Hz transformer PSUs

The simpler type of PSU fitted to most homebase VCRs, many monochrome and small-screen TV sets, and much other equipment is represented by the diagram of Figure 3.15. It gives mains isolation in a simple way, at the cost of the weight and expense of the mains transformer and the size and finite life of the large electrolytic capacitors required. The mains transformer secondary winding may have a centre-tap as shown here, or take the form of a simple winding feeding a bridge rectifier – rectification is invariably full-wave in these designs.

Faults in this type of circuit are not difficult to diagnose. Leakage or short-circuit in the series regulator element TR1 causes excessive output voltage accompanied by hum-modulation of the line at 100Hz rate to give a double (see Figure 3.13) bar on the picture. The same 'hum' effect can be due to failure of the reservoir capacitor C1 in Figure 3.15, but in this case the output voltage *falls* rather than rises. Where the ripple (as shown on TV screen or by the scope) is at 50Hz, the circuit has undoubtedly gone into half-wave mode. Ripple frequencies can be difficult to read quickly on a 'scope – compare with true mains frequency on the second trace of a dual-channel scope. Merely holding no. 2 probe tip in the fingers will suffice.

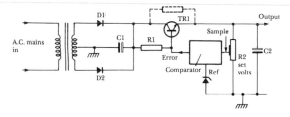

Figure 3.15   *The series regulator – a simple form of voltage stabilizer*

Half-wave operation may be due to failure of one of the rectifier diodes – or in the case of a centre-tapped transformer secondary, failure of the winding itself or the connections to it. If the design of the reservoir and regulator circuit is good, the half-wave rectification may not have any noticeable effect on the output line voltage or ripple level, though the mains transformer may heat up and generate a strong magnetic field when one half of its secondary winding is loaded in this way. In one example encountered on the bench, the situation was revealed by a moving coloured pattern in one corner of the TV screen, the effect of a strong non-synchronous 50Hz magnetic field on beam landing/purity of the picture tube adjacent to where the transformer was mounted; removal of the back cover revealed a very hot transformer and an open-circuit connection to one of the two rectifier diodes.

Sometimes, some of the burden is taken from the regulator transistor by a parallel resistor, shown dotted in Figure 3.15. If this goes open-circuit, TR1 may overheat whilst giving a shortfall in output voltage. Similarly, if the regulator turns off, some output will be maintained by the (overheating) parallel resistor.

In cases of low output voltage from a circuit of this type, it's important to remember that when the collector voltage falls to a point near the normal output voltage, collector current will cease and the load will attempt to draw on the transistor's base current supply, pulling that low. Before investigating the regulator drive circuits, then, ensure that collector voltage is maintained at a suitably high level, e.g. by a good reservoir capacitor. Other causes of incorrect output voltage, jitter etc. are similar to those already described for SMPSUs.

In high-current 50Hz and 100Hz power supplies, typically thyristor-controlled types, a circuit similar to the regulator part of Figure 3.15 is used not as a voltage stabilizer, but as an active ripple-filter. Its 'stabilization' action is effected at mains rate. Sometimes a d.c. feedback circuit is built round it to set up the desired PSU output impedance. Any problems here are usually manifest as hum-ripple on the output line, and due to faulty semiconductors in the ripple-filter circuits.

IC-based series regulator systems

Series-regulator systems often come packaged in IC form, ranging from the fixed-voltage 3-leg device

shown as IC621 in Figure 3.6 to the more elaborate but basically similar stabilized supply package of IC151 in Figure 3.16. All these incorporate feedback, reference and driver circuits and have provision for heat-sinking. The main point to bear in mind when servicing is the influence of the various control pins (nos 2, 11 and 13 in Figure 3.16) one or more of which are governed (especially in the case of a VCR) by the system-control section. The chip should not be condemned until it's proved that:

1 Adequate supply voltage is present at the input.
2 The load on the output is not excessive.
3 The control pin, where relevant, is commanding 'on'.
4 The ground pin of the IC is truly grounded to PSU and load earth.

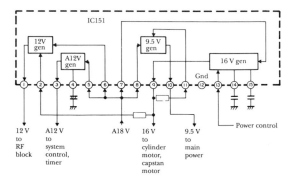

Figure 3.16 *Multiple series regulators in a heat-sinked IC package for VCR use (Hitachi)*

References

VCRs often have small d.c. – d.c. and filament-voltage convertors to power fluorescent display panels – mention of these is made in Chapter 19.

Some switch-mode power supplies use the chopper transformer as a source of focus and EHT voltage via diode-split overwindings. These are discussed in connection with line output transistors in Chapter 4.

PSUs of all sorts often suffer from intermittent faults – their causes and effects are described in Chapter 20.

Practical repair hints are given in Chapter 21, and Appendix 1 (safety) is very relevant to PSU servicing work.

4

LINE TIMEBASE

If the line timebase (LTB) were purely that, it would not merit as long a chapter as this will be. While the basic function of the line timebase section is to deflect the scanning spot across the face of the picture tube, it can have many auxiliary roles: generator of accelerating, focus and EHT voltages for the picture tube; modulator of scan-error-correction influences onto the horizontal scanning waveform for some types of picture tube; distributor of line-rate pulses to many parts of the set; source of low-voltage supplies for audio, field timebase and signal stages in some designs of TV; provider of heater energy for the picture tube; synchronization source for the PSU section; and in some designs an integral and inseparable part of the PSU itself. Most of these auxiliary functions stem from the fact that there must necessarily be a line output transformer (LOPT) to provide the essential scan-current-shaping inductor, and to match the impedance of the line scanning coils to that of the line output stage. A transformer alive with pulses at 15-odd kHz is a useful and efficient source of secondary energy. Even so, the fact that most PSUs now operate at similar frequencies to LTB's and with similar inductors, means that the 'universal provider' role is increasingly being shifted to the PSU proper – indeed some LOPTs in modern TV designs are matchbox-size 3- or 4-pin devices: here all auxiliary supplies come from the chopper transformer.

Because we are taking a broad view of LTBs, and purely from a servicing point of view, we shall examine all its possible roles, but bear in mind that progress has (as in all areas of TV design) changed circuit designs considerably over the years. Many types of picture tube do not require pincushion correction – it is built into the deflection yoke. Many sets appearing on the repair bench *do* have raster-correction circuits, and they are a common cause of trouble: reference to them will be found here. Similarly, most modern LOPTs have internal supply networks for 1st anode, focus anode and EHT supplies, with built-on trimming controls – as well as covering these, we shall regard the older and more troublesome EHT tripler assemblies and potential divider networks whose breakdown can and does cause trouble.

LTB block diagram

A block diagram of the line timebase section of a TV set is shown in Figure 4.1. Its primary function is to deflect the scanning spot in the picture from left to right of the picture-tube faceplate at $64\mu s$ intervals. The line timebase is broadly divided into three sections: an oscillator capable of being synchronized to incoming line pulses; a driver stage to provide an impedance match between oscillator and output sections; and the output stage itself, whose load is the scanning yoke. The key waveforms are shown in the diagram. The output from the oscillator section is a square wave with $64\mu s$ period, and the waveform at the collector of the driver

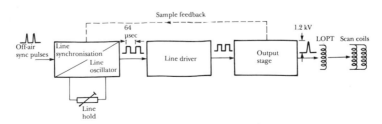

Figure 4.1 *Line timebase block diagram – each section is described individually in the text*

33

stage is an amplified and inverted version of this. The voltage waveforms in the line output stage all have the appearance of a single flyback pulse per $64\mu s$ whose amplitude at the collector of the line output transistor is typically 1200V, and whose duration is about $11\mu s$. These amplitude and duration characteristics are important, as will become clear. The voltage waveform gives no indication of the current flowing in the inductive components of the line output stage. This current is an almost pure sawtooth, visible by measuring the voltage developed across a low-value (less than 1Ω) sampling resistor in series with the inductor, be it the LOPT or scan yoke. In practice, it is seldom necessary to do this; interpretation of the flyback voltage pulse is usually sufficient for fault diagnosis.

Line oscillator section

Most line oscillators are embodied in integrated circuits and typical set-up is illustrated in Figure

4.2. Correct free-running speed of the line oscillator is essential, and every oscillator has timing components (usually C/R, sometimes C/L) to determine basic frequency. Here the frequency is controlled by the network C752, R754 and trimmed by RV741. Any large departure from normal frequency, indicated by line break-up on screen or in extreme cases by timing indication on a scope, should direct attention to this R/C network, though the influence of the phase control section within the chip must be borne in mind. This particular line oscillator runs at twice speed, 31.25kHz, for reasons of automatic field sync, but a ÷ 2 stage is interposed between the oscillator and the chip's output pin 10.

A complete loss of action in the set results from a stopped line oscillator. If in the circuit of Fig. 4.2 there is no output at pin 10, the first check should be for presence of 12V supply at IC pin 16, and via R757 at pin 10 itself. If these are in order and the R/C network mentioned above is OK, the chip itself is implicated. When checking supply voltages remember that some sets (especially older designs)

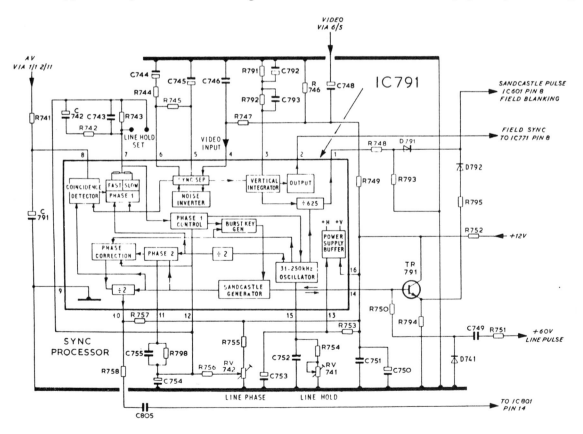

Figure 4.2 *Line oscillator and sync section using a TDA2576 chip (Ferguson)*

have a kick-start system to get the line oscillator going, after which the supply is sustained from the line output or chopper stage. For diagnostic purposes the 'start' condition can be maintained by bypassing the kick-start charging capacitor with a heavy wire-wound resistor (typically 5.6kΩ 10W, check for correct operating voltage at chip) or preferably by application of a suitable voltage from an external power supply unit, or even a battery.

However, most problems that assail the line oscillator section have to do with synchronization, jitter or phase rather than complete failure. In many designs (though not in that of Figure 4.2) there is a high-value (i.e. greater than 100kΩ) resistor associated with the pin through which the video signal passes on its way to the sync separator within the chip. This biases the separator stage, and weak, intermittent or picture-dependent sync is often due to failure of this resistor. Where such a resistor is not fitted, or is OK, it's worth checking the quality and amplitude of the sync pulse train itself (here at C748) before suspecting the chip itself – the pulses can be degraded by an i.f. or post-detector video fault: see Chapters 6 and 7.

Most of the other sections of the chip are concerned with synchronization of the oscillator section, and its pulse phasing relative to the video signal. The phasing of the line drive pulse determines the horizontal position of the picture on the screen. This is adjustable to some degree with the phasing control RV742. A large horizontal error in picture position should lead to a check of this, also C754 and R798. This particular circuit primarily achieves synchronization internally, in that the phase control 1 section compares the timing of incoming sync pulses with those of the oscillator; a second phasing influence comes into pin 14 from the LOPT to compensate for phase shifts in driver and output stages. If the picture moves horizontally with changes in brightness and contrast, this feedback is suspect – check the components involved, here D741, C749 and R751.

Horizontal jitter or 'twitch' can be a difficult fault to deal with, and can spring from several sources. Where an IC is used as line oscillator, the chip itself is not as often responsible as its peripheral components: those in the flywheel time-constant section (here C742, C743, etc.; C744, C745 are also worth checking); and particularly supply-line decoupling capacitors: C750 and C753 in this circuit. Picture 'topwobble' in which the upper part of the picture is hooked at an angle, or waves to and fro, usually stems from a problem in the flywheel filter (e.g. C742/3 in Figure 4.2) or a fault in the sync separator section. Where (as is most common) it occurs on VCR playback, the cause is that the flywheel circuit cannot cope with the timing jitter impressed on the playback signals. This may stem from mechanical faults on the videotape deck (see Chapter 12) or from an incorrect time constant in the flywheel smoothing circuit of the TV. Some very old sets need modifications; most modern types can cope with VCR playback timing jitter, but it may be necessary to select a 'reserved' channel (typically 0, 6,8 or 16; see TV instruction book) to invoke the correct (shorter) flywheel time constant and thus prevent 'watery' verticals and hooking at the picture top. Figure 12.14 shows the effect. In Figure 4.2 selection is made by an AV switching line from the channel-selector section which raises IC791 pin 8 from 1V to 3.5V to invoke 'fast' mode in the 'phase 1' block.

The sandcastle pulse generated within IC791 has its counterpart in most modern TV designs. In general, it has three components:

1 A field blanking pulse at 20ms intervals which usually embraces the entire field blanking period to suppress bright dots at the picture top due to Teletext pulses.
2 A line blanking pulse of about 12μs duration at the same amplitude.
3 Sitting on top of the latter, a burst gating pulse timed to coincide with the colour burst in the transmission, for use in the decoder chip. Problems like no colour, no picture, incorrect blanking or wrong luminance levels, then, can stem from a defective sandcastle pulse or incorrect phasing of the line oscillator: the timing, amplitude, shape and d.c. level of the sandcastle pulse all merit careful study where these symptoms are encountered.

The line drive emerges from pin 10 of the chip in Fig. 4.2. Normally it feeds a line driver stage (see later) but sometimes, as here, it merely synchronizes the PSU control chip, whose output pulse is eventually applied as drive to the line output stage.

Alternative IC line oscillator

A simpler form of IC line oscillator, representative of the older designs of TV which find their way to the work-bench, is shown in Figure 4.3. The

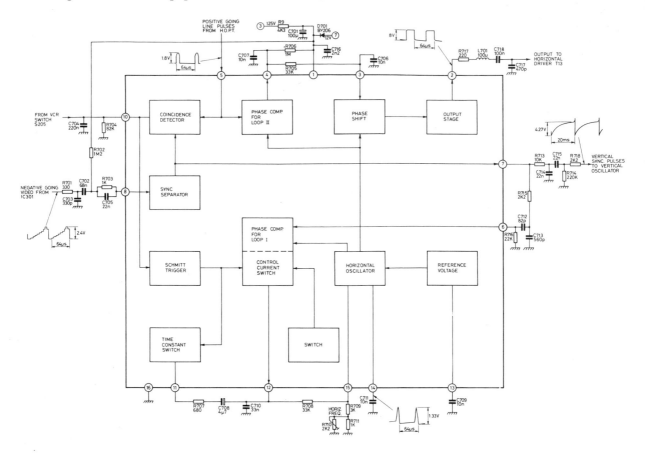

Figure 4.3   *Early form of IC-based line generator using TBA920 (ITT)*

TBA920 is a well-used chip, and all manufacturer's circuits using this device are necessarily similar. Many of the remarks made above apply equally to this circuit. The reference component for line frequency is C711 – change in value or S/C will result in incorrect line speed and no oscillation respectively. The circuit has no line phase control because most TV sets using it had a separate line shift control in the line output stage; see later. However, faults in C707 or associated components can cause horizontal displacement of the picture, as can incorrect timing of the LOPT-derived pulse applied to IC pin 5 for comparison purposes.

Composite video enters R701 for sync separation; loss of sync may be due to lack of video here, and poor sync to a faulty 1M2 bias resistor R702. VCR time-constant switching is applied to pin 10 of the chip which is grounded for short time-constant VCR operation. If this pin is grounded during normal (off-air) reception, a 'ragged' picture results if the received signal is less than perfect.

Discrete line oscillator

As a final illustration of line oscillator types and diagnosis procedures, Figure 4.4 shows a form of discrete-component design found in many monochrome TVs and some older colour types. This is a sine-wave oscillator, distinguishable by the fact that the line hold control is generally a variable inductor rather than a resistor, though a resistive preset may be incorporated, as here. The oscillator itself is VT502; VT501 behaves as a variable tuning capacitor under the influence of its base bias voltage, derived as an error signal from the phase detector W501 and W502.

A complete lack of oscillation, where HT supply

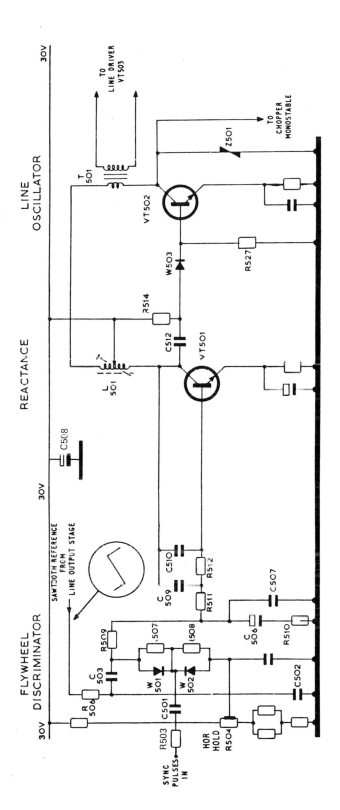

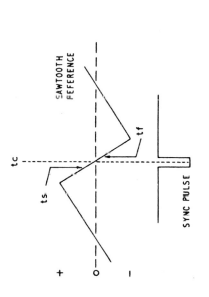

Figure 4.4 *Sine-wave oscillator with reactance control of frequency for flywheel synchronization (Ferguson)*

is present at the tap on L501, may well be due to failure of decoupler C508, though a faulty oscillator transistor VT502 is a strong possibility. While problems in the reactance stage VT501 etc. can stall the oscillator, they are more likely to cause loss of line hold due to the oscillator running at the wrong speed. Anything that changes the effective capacitance at the bottom end of L501 has this effect, so VT501 and the four capacitors associated with it are suspects. More often the error voltage developed across C507 is incorrect. If the manufacturer quotes a voltage for this point, a quick check is to inject that voltage from a low-impedance external source: if line frequency is then wildly wrong the problem is in the oscillator or reactance stages, whereas a return to about normal frequency indicates a problem in the phase detector section. W501 and W502 are the first components to check, then any electrolytic capacitors – here C506.

In this type of circuit, weak or absent line sync (where the oscillator will 'run through' but not phase-lock to the transmission) may be due to loss of *either* the incoming sync at R503, or the feedback pulse from the LOPT, here coming via R506. Incorrect line phasing (see secondary diagram below) will be due to wrong timing of the feedback pulse – check R506, C502 and upstream components. Some circuits of this type have two (antiphase) feeds from the LOPT, in which case check both sets of integrating components.

Line driver stage

The driver is one of the more reliable components of the line timebase; except where the line output transistor is a Darlington type (i.e. BU807 or similar) the driver consists of a medium-power transistor with a step-down transformer in its collector circuit. The transformer secondary winding feeds the base of the line output transistor. To prevent overheating of the line output transistor the drive (checked at the driver collector) should be a square-wave with peak-to-peak amplitude approaching that of the driver's supply voltage. If not, or if driver performance is erratic, check any supply decoupling capacitors, and (if fitted) the damping components across the primary winding of the transformer. Failure of the latter can lead to breakdown (sometimes sporadic) of the driver transistor. The joints on the driver transformer are also worth checking in cases of intermittency.

Line output stage

A skeleton circuit of the line output stage is given in Figure 4.5, where the line output transistor is shown as a switch – in practice it is exactly that, and it is important for correct operation that it switches cleanly, quickly and fully on and off. This depends on its drive waveform – anything less than a full-amplitude square wave will lead to a high dissipation (hence overheating and failure) in the line output transistor. Since the base drive is essentially a current one, voltage oscillograms at the base are misleading. Check any series resistor/diode network once it's established that the driver collector waveform is full and square.

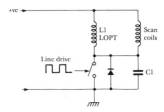

Figure 4.5 *Basic principle of operation of line output stage*

Each time the switch closes, the bottom end of LOPT L1 is grounded, resulting in a linear build-up of current in this and the scan yoke. The amplitude of the current ramp is proportional to supply voltage, so correct and constant supply voltage ensures full and stable picture width. At the end of scan the output transistor switches off, releasing the bottom of L1 from ground. L1's top end is still clamped to supply voltage, so its stored energy now produces a positive half-cycle of oscillation at the collector of the now open-circuit switch transistor. This flyback pulse, height and width-wise, is governed by the LC circuit formed by L1 and C1, respectively the LOPT and the tuning capacitor. If the tuning capacitor (always connected across the line output transistor) should go low capacitance or O/C, the flyback pulse becomes thinner and taller, creating dangerously high EHT and collector-pulse voltages. Symptoms of these are a small bright picture at normal HT line voltage, and premature failure of the LOPT or line output transistor due to insulation breakdown. Where this situation is suspected, restriction of the supply voltage will permit safe testing of the flyback period with a 'scope: it should be around 11 μs. Even with a very low applied voltage (i.e. *tens* of volts) the test for flyback period is valid since it depends only on L/C constants. Ensure that any replacement tuning capacitor is the correct type – frequency and ripple current ratings are important.

When the width of the flyback pulse is right its amplitude (so long as HT supply voltage is correct) should also be right – 1.2kV at the transistor collector is a typical amplitude for a colour TV. An oscillogram of a correct line flyback pulse is shown in Figure 4.6.

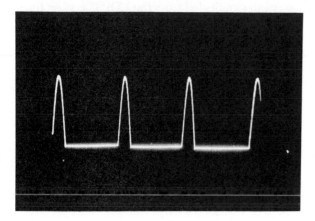

Figure 4.6 *Line flyback pulses as they appear at the pins of the LOPT*

Our first example of a complete line output stage is given in Figure 4.7. This oldish design is an ideal model for demonstration of fault patterns and diagnostic procedures, most of which are applicable in general terms to the later models to be dealt with shortly. The most common fault symptom, perhaps, in a line output stage is complete lack of operation, with the PSU pumping or a supply fuse blown. In these circumstances the first check should be across the line output transistor (here Tr401) for S/C or heavy leakage. This simple test, easily made from the collector (metal case or heat-tab) of the device to chassis or 'ground' line, should read low one way (diode conducting) and high the other. If low resistance in both directions is read, first isolate the supply (here, 148V) rectifier at the PSU because it is effectively in parallel with the line output transistor. If the low reading remains, isolate and check the line output transistor Tr401, parallel efficiency diode(s) D401/2, S-correction capacitor C508 etc. On rare occasions the tuning capacitor (here represented by C403/4/2), the LOPT, or decoupler C405 will be found responsible.

Very often the resistance check described will show nothing wrong, but the stage refuses to work and attempts to draw a heavy current. This indicates a heavily loaded LOPT whose inductance is effectively damped by parallel resistance, and in

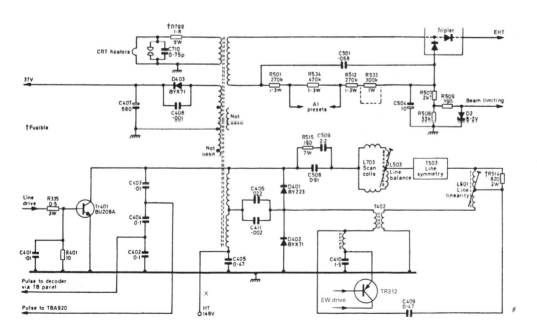

Figure 4.7 *Line output stage for use with thick-neck (20AX) picture tube (Decca)*

many designs gives a 'dead-set' symptom, with no picture or sound. The way to deal with this is to progressively unload the peripheral circuits from the LOPT, checking each time whether the loading problem has disappeared.

In the circuit of Figure 4.7 the first move would be to disconnect the tripler from the LOPT output nipple. Keep the lead-end well clear of the LOPT to avoid sparking. If normal operation is now restored (i.e., tube heaters light up, see top left of diagram) the tripler is suspect, but always check the overwind's bottom-end pulse decoupler (here C501) before condemning the tripler, and ensure that any replacement is correctly wired, especially regarding 'diode' and 'capacitor' leadouts. If the

loading remains with the tripler off, the next steps would be disconnection in turn of D403, R709, C406/411, and finally the line scan coils.

Figure 4.8 shows a different design of line output stage to highlight other possible causes of LOPT loading: chokes and transformers used for coupling and pulse-isolation of *line-shift circuits* (i.e. L30) commonly go S/C turns; *E-W modulator diodes*, often used to provide LT supplies too, D23/24; *transductors* used for raster-geometry correction, L5/6 and associated components; and S/C or leakage in circuits fed from the LOPT, e.g. on lines *V, VI* or the pulse feeds from L22/3/4. Typical of the latter would be a dead-shorted field output IC pulling down its supply lines. Disconnecting the

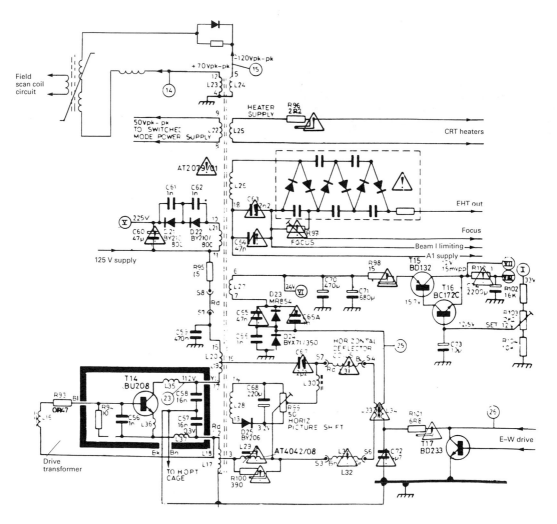

Figure 4.8 *'Balanced' line output stage, with the LOPT primary winding split into two sections (ITT)*

LOPT windings will isolate these problems for diagnostic purposes.

When we arrive at the point where the LOPT stands alone with only its primary winding in circuit, the likelihood is that the LOPT itself is faulty, with S/C turns. Passive tests of a LOPT with an ohmmeter are seldom conclusive; 'pulse' testers are available commercially, see p. 17. For diagnosis of problems in heavily loaded line output stages the applied voltage can be reduced to a safe level by either substituting an externally derived low voltage supply for the 'real' one, or (more practical) by limiting the current available by means of a domestic 240V light bulb in series with the supply – point X in Figure 4.7, or in place of R95 in Figure 4.8. An oscillogram of the waveform at the collector of a line output transistor is shown in Figure 4.9: the LOPT is faulty with S/C turns, and the current was limited by use of a series bulb in the supply line. Further evidence of the source

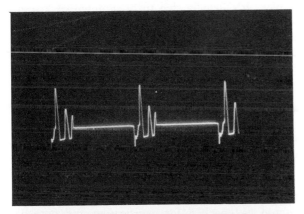

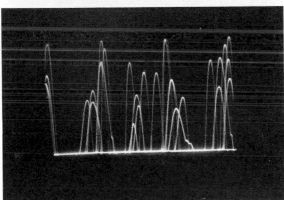

Figure 4.9 *Oscilloscope trace from a LOPT with short-circuit turns. Two examples are shown here: in (b) the line frequency has been upset by the faulty transformer*

of the problem was that the LOPT became excessively warm after a short period of operation. If a LOPT is proved faulty (and especially where the problem seems to stem from overheating in the 'overwind' section) bear in mind the possibility that a faulty tripler may have caused its failure. This does happen in some designs, in spite of the presence of over-current trips! Check the new LOPT first with the tripler disconnected, and switch off quickly if loading is apparent with the old tripler reconnected.

Failure of line output transistor

After replacement of a faulty line output transistor, it may be found that the new one fails at switch-on or after some period of running. Possible causes for this are:

1 Unsuitable 'equivalent' type fitted.
2 Faulty tuning capacitor, i.e. C57/C58 in Figure 4.8.
3 Dry-joints at LOPT or sparking/arcing at tripler, focus pot, scan yoke connectors, etc.
4 Intermittent over-voltage from PSU, due typically to a dry-jointed feedback winding, sparking at main reservoir capacitor, or failure of regulator section.
5 Flashover in the picture tube.
6 In some chassis, poor or intermittent contact at the mains outlet, i.e. via a latchety multi-way adaptor or loose fuse in mains plug.
7 Weak or 'lazy' line drive; see 'line driver' section above.

Fault symptoms

Because EHT voltage and/or tube heater current are very often derived from the LOPT, most faults in the line timebase result in no picture at all, as has been described above. The operation of the heaters can be checked visually in most tube types, and the presence of EHT is indicated by a 'bristling' feeling on the back of a hand held in front of the tube's faceplate. A rustling noise at switch-on is further evidence of EHT voltage presence.

Less serious faults in the line timebase do permit the display of on-screen symptoms, described with possible causes below.

Vertical line only down screen centre

In a conventional line output stage this can only stem from an O/C in the yoke circuit itself. Check for dry-joints, bad plug/socket connections, burn-ups in convergence components, faulty S-correction capacitors etc. The problem is very seldom in the scan yoke itself. To illustrate the possibilities, the yoke circuit in Figure 4.7 is LOPT, C508 (S-correction), L703, L503, T503, L501, T402, C405, LOPT. In Figure 4.8 it is LOPT pin 16, C67 (S-correction), socket S7, yoke L31, socket S4, L33, socket S6, yoke L32, socket S3, L29, LOPT pin 3. Long and vulnerable paths!

Where the EHT etc. are sourced from a chopper transformer, any part of the 'real' line output stage may be responsible – check with scope and meter.

Narrow picture

There are several possible causes for this. First check HT voltage at the 'top' of the LOPT primary winding. If it's correct, check the period of the flyback pulse: less than $11\mu s$ throws suspicion on the tuning capacitor. If the flyback period is correct, check any diode-modulator drive circuit: these are only fitted to sets using tubes which require pincushion correction, and incorporate resistive width controls. If the width pre-set does nothing, the fault will be (electrically) very near the pre-set itself.

Excessive width

Where the picture width, according to a test-card, is excessive, the first step is again to check the HT voltage – width is proportional to supply voltage. If HT is correct, the two possibilities are that the E-W diode modulator or its drive transistor/circuit are in trouble – see below and try the effect of the width control – or that the LOPT tuning is incorrect: measure flyback pulse width as already described. As common a failing in this part of the circuit is the following.

E-W pincushion distortion

This effect, illustrated in Figure 4.10, is usually accompanied by a lack of response to the pre-set controls for width, pincushion and trapezium

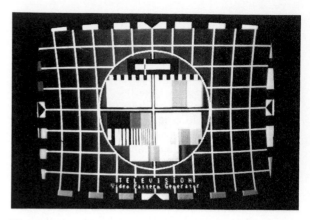

Figure 4.10 *The effect of E-W (pincushion) distortion on the picture*

correction; and sometimes develops at the time of failure of some LOPT-associated component like the line output transistor, focus unit or EHT tripler. It's usually due to breakdown of a semiconductor, quickly identified in a 'cold' check with an ohmmeter. Primary suspects for this and the 'overscan' symptom outlined above are the EW modulator diodes and the transistor which drives them: D401/2 and Tr312 in Figure 4.7, D23/4 and T17 in Figure 4.8. As a quick check, picture width should flick from one extreme to the other when the driver transistor is effectively opened and shorted c-e. If not, check the diodes and circuit continuity.

Where semiconductor failure has occurred, the wound components immediately associated with them (EW loading coil and coupling transformer) should be checked for overheating and damage. If replacement components – wound or semiconductor – overheat, check the transistor's parallel capacitor: here C410 and C72 in figures 4.7 and 4.8 respectively. The E-W output transistor dissipates considerable energy and needs to be well heatsinked.

Picture ballooning

If the picture changes in size as brightness and picture-content varies the cause is poor regulation of supply lines. Check that HT supply voltage stays reasonably constant, i.e. within 3%. If not, concentrate on the PSU, checking reservoir capacitors and feedback circuit. If HT level is constant, suspect the EHT tripler and check for any excessive 'series' resistance in the LOPT primary circuit depicted in Figure 4.5, and in the EHT generator circuit, not forgetting the link between the lower end of the overwind and ground. Some EHT connectors (at the tube bowl) incorporate series resistors whose failure can cause picture ballooning. This is often given away by signs of overheating or swelling (and sometimes a nasty smell) at the connector stem. In rare cases, intermittent contact and sparking at this point can invoke PSU tripping.

Discharge effects

Corona and spark discharge can occur at connection points, spark-gaps and elsewhere in EHT, focus and (to a lesser degree) A1 supply circuits. Careful study of the screen can help pinpoint the exact trouble spot: a vertical 'brushing' line on picture (Figure 4.11) indicates a corona discharge at some a.c. (pulse) point, identifiable by a smell of ozone and a blue ionization glow (view in darkness) at the trouble-spot in the set. On the other hand, *d.c.* discharges may quietly give rise to white spots all over the picture area, or to sudden loud and frightening cracks which at best trip the PSU, and at worst destroy one or more semiconductors in the set – possibly as far away as the field timebase, Teletext decoder or channel-select sections. The problem is to identify the exact point of discharge. If the focus unit is connected directly to the EHT supply it is suspect. Alternatively, signs of discharge may be visible: black streak-marks on the lip of a diode-split LOPT's output socket or a pinhole in the plastic body of an EHT rectifier or tripler. Corresponding marks may be found on adjacent metalwork.

Intermittent discharge in the focus spark-gap or nearby can have several effects: a 'snicking' noise audible in quiet surroundings; white defocused blobs on the picture; momentary blackout or defocusing of a single spot or line of the picture; intermittent frame roll, triggered by each spark; or intermittent PSU tripping. When quiescent, this fault can often be provoked by turning the focus control to full voltage, then if necessary blowing moist breath(!) on the spark gap and its surroundings. For other focus problems see Chapter 10.

Ripple on auxiliary supply lines

Where auxiliary supplies are derived by rectification of flyback or scan pulses from the LOPT, insufficient smoothing can impart a line-rate ripple to the supply line. This may typically show up as shading on the picture where a line-rate sawtooth is present on (e.g.) the 200V RGB amplifier supply line or the d.c. feed to the picture tube first anode(s). A typical effect is shown in Figure 4.12, where the reservoir capacitor for the RGB amplifier's operating voltage (C60 in Figure 4.8) is O/C. The same applies, of course, to supplies derived from a chopper transformer, and in either case a scope check on the line shows the ripple superimposed on the d.c. voltage.

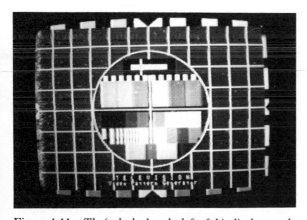

Figure 4.11 *The 'splashes' at the left of this display are due to a discharge in the line output stage or some associated component working at high pulse voltage*

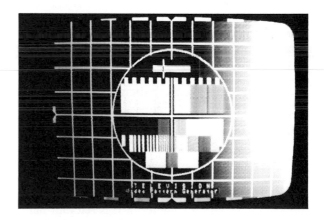

Figure 4.12 *The shading effect shown here indicates line-rate brightness modulation of the video signal*

Geometric distortion

Raster shape distortion can take many forms. A
wasp-waisted or keystone-shaped raster suggests
that the HT supply line is carrying ripple. A
keystone effect with *convergence errors* is a sure sign
that one-half of the scan yoke is faulty, badly-
jointed or incorrectly fed. In circuits using diode
modulators (i.e. Figures 4.7 and 4.8) there is a
keystone control pre-set, so mild keystone distortion
without misconvergence should lead to a check on
its setting and the field-rate sawtooth waveform(s)
fed to it.

In older sets where transductors (saturable transformers) are used to cross-modulate line and field
scanning waveforms, the transductor itself and
associated R/C components are suspect for
poor raster shape which cannot be improved by
adjustment.

Poor horizontal linearity is most likely to be
caused by a fault in the S-correction capacitor (C67
in Figure 4.8 or one of the network C508/9, R515 in
Figure 4.7) or a need for adjustment/replacement of
the line linearity controls L29, L501 respectively.

Vertical striations

Each of the linearity control coils mentioned above
is damped by a parallel resistor, R100 and R514
respectively. If the resistor should go O/C the coil
will ring at each flyback pulse to velocity-modulate
the tube's electron-beam and give vertical bars at
the LHS of the screen, becoming fainter towards
the centre. The same effect can be caused by
intensity modulation of the beams, the result of a
damped oscillation being fed to the tube's cathodes
or grid. This may be caused by a faulty damping or
blanking circuit, failure of a decoupling capacitor,
or even poor dressing of the wiring harness. The
first step is to find it with an oscilloscope, working
on a 'blank raster' from a pattern generator.

Beam-current limiting

At some point the beam current in the picture tube
is monitored for the purpose of limiting it to a safe
level. While faults in this area are covered in the
luminance section of Chapter 6, it is relevant here
to examine some of the causes of failure of beam-
limit circuit components, and of false high-current
indications. In most sets the beam current is sensed

via the components in the 'bottom-end' of the
overwind/tripler circuit. D2 in Figure 4.7 and the
junction of C63 and the focus control in Figure 4.8
are the take-off points. A faulty tripler or problems
in the R/C network can falsely signal excessive
beam current, turning down brightness, contrast or
both. The same causes – or flashover – can
permanently damage semiconductors in the beam
current limiter circuit; overriding the BCL sensor
with an externally derived voltage (or grounding it,
see manufacturer's voltage data) will quickly prove
the point. Failure of a smoothing capacitor in the
beam current limiter circuit can give a shading
effect, similar to that of Fig. 4.12, to the picture,
especially in early TV designs where beam current
is sensed across a resistor in series with the line
output transistor.

Whistling components

A TV set contains many ferrite-cored components
handling considerable energy at 15kHz or similar
frequencies: the LOPT, chopper transformer, scan
yoke, driver transformers, line linearity coil, etc.
The effect of magnetostriction can give rise to an
audible whistle from any of these. Where the
offender is a large component with no clamps to
tighten, it will have to be replaced. Small components, and in particular line linearity coils (the
worst offenders) can often be silenced by thoroughly
anointing them with white resin wood glue, or hot-
melt adhesive from a glue gun.

Later timebase circuit

Figure 4.13 shows the circuit of a more modern line
output stage. Because most late-type picture tubes
do not require E-W raster correction (it is built into
the yoke design) provision for this is optional,
involving removing link Lk33 and inserting a
modulator diode and associated drive components,
similar to those we have already seen. In other
respects the circuit corresponds to those we have
already examined (though this one is much simpler)
and all the same remarks apply. Much of the
simplification comes from the use of a diode-split
LOPT, in which the diode and capacitor components
of the tripler are built into the LOPT overwinding,
as are the focus and A1 voltage supply networks
and control pots. In the event of a fault developing
in these internal components, the LOPT will 'load'

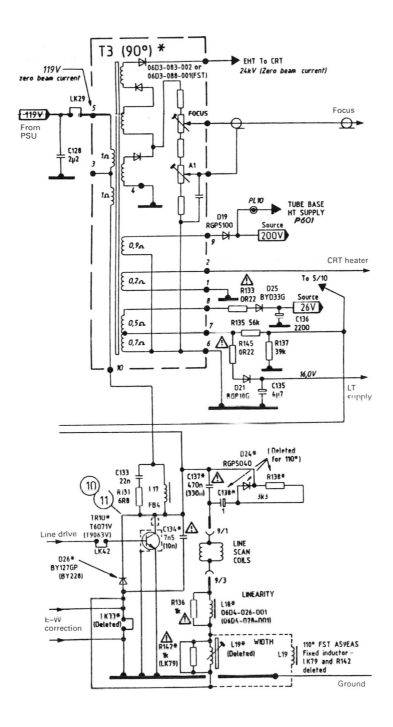

Figure 4.13   *A later example of a line output stage (Ferguson)*

as if it had S/C turns, and a method of diagnosis for this has already been suggested. Some small monochrome LOPTs have an EHT rectifier stick moulded into the overwind section, and this can fail to heavily load the transformer. In this case, disconnecting the tube final anode connector will remove the capacitive load of the tube's conductive coatings, permitting the LOPT to work again. A (temporary?) repair of sorts can be made by wiring an external and heavily insulated EHT stick in series with the lead to the tube cavity connector.

The d.c. resistances given for individual LOPT windings on this diagram bears out the earlier suggestion that S/C turns cannot be detected with an ohmmeter!

The separate inductive width control L19 represents a method of adjusting scan amplitude when no diode modulator is present, and is shunted by a resistor R142 whose failure (O/C) could cause vertical striations down the LHS of the screen. Most designs do not have this type of width control, relying instead on close tolerance components and judicious setting of the PSU output voltage control. The linearity correction coil here, L18, is not adjustable.

Many of the components whose possible failure and detection were discussed earlier can be identified in Figure 4.13. The tuning capacitor is C134, S-correction/yoke coupler C137, and primary decoupler C128. Several auxiliary supply lines are fed from secondary windings on LOPT T3, but the tendency for manufacturers to fit fusible resistors (R133, R145, etc.) means that they will open in the event of excessive loading, releasing the LOPT and simplifying the diagnosis.

There are many variations in line timebase circuit design and construction over the range of TV sets encountered on the repair bench. In virtually all of them the component types, functions and failure modes described so far in this chapter are applicable. Even the very tiny, low-power line timebases used in tube-type TV cameras for vidicon scanning, and in electronic viewfinders for horizontal deflection and generation of accelerating voltage, work to the principles in Figures 4.1 and 4.5; show oscillograms like figures 4.6 and 4.9; can give rise to faults like those of Figures 4.11 and 4.12; and have wound components which can 'ring' to give picture striations or audibly whistle as described above.

The only exception to the rule is the *thyristor* line timebase to be discussed next.

Thyristor line timebases

Although the thyristor line output stage has not been incorporated in new designs for some years, enough sets using this configuration are still in use to justify a brief description here.

Fig. 4.14 shows a skeleton circuit of a thyristor output stage. For a description of its operating principles, see *TV and Video Engineers Pocket Book* (Heinemann); here we'll confine ourselves to fault symptoms, causes and their diagnosis.

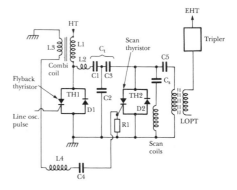

Figure 4.14 *Skeleton circuit of thyristor output stage*

The wound components used in those circuits are physically heavy, and handle large currents. These two factors lead to a high incidence of dry joints on the combi choke, LOPT etc., and these should be checked first in all cases of trouble: typical symptoms are PSU pumping, cut-out operated, burning and 'no-go'.

Where dry joints are not responsible, the cause of excessive current consumption can be quickly isolated by shorting the anode and cathode of the scan thyristor with a wire link. The set can now be safely switched on. If the excessive current (i.e. PSU tripping) continues, the problem is in the area of the flyback thyristor: check the thyristor, its drive pulse condition, the parallel diode and any loading effects on the combi coil, i.e. secondary winding/rectifiers, C1 etc. If, however, the loading effect disappears with the scan thyristor shorted, the LOPT/scan yoke circuit is implicated: check the tripler first by disconnecting it, then progressively unload the LOPT as already described for conventional line output stages.

A 'dead' line output stage drawing little or no current from the power supply may be due to an O/C in the flyback thyristor TH1 or its lack of triggering pulses; or a S/C in the scan section, e.g. TH2, D2, C5 etc.

Other faults possible in thyristor time timebases are as follows:

1 Incorrect width, foldover, picture ripple – broken or cracked ferrite components.
2 Incorrect width, jagged verticals, 'sawtooth' effect on picture – faulty transistor/diode/transformer in width modulator circuit where this is part of the line output stage and in parallel with the input coil L1 in Figure 4.14.
3 Foldover/distortion at picture centre – incorrect timing of flyback trigger pulse due to faulty feed choke etc. at TH1 gate.
4 When combined with low width – efficiency diode D2 O/C.
5 Foldover at left of picture – O/C flyback diode D1.
6 Loss of auxiliary supplies/no start-up – faulty diode on combi choke secondary winding in some designs. Check any fuse here first, however.

Some thyristor line timebase designs depend for EHT and picture width stabilization on a 'dumping' circuit based on a third thyristor in the PSU section. When energy level in the thyristor line output stage reaches the correct point for operation this device comes into operation to pass excess energy into the PSU reservoir capacitor. If it fails to fire, some form of overvolts trip will come into operation to shut the line output stage down. The symptom is a brief burst of energy at switch-on. In these circumstances, the best way to approach the job is by use of a mains variac. Wind up the mains to the point where the set is operating correctly, then check with oscilloscope and meter the operation of the *set EHT* preset and the regulator circuit generally.

Replacement components for the LTB section

In relative terms the components in LTB stages (and PSU sections, indeed) are heavily loaded. High currents, high flyback voltage peaks and the fast rate of energy exchange involved in scanning at 15.625kHz rate combine to make considerable demands on components, especially critical ones like the line output transistor, tuning capacitor, yoke coupler/S-correction capacitor etc. For safety, reliability and correct performance it is important that any replacement components be acceptable in several respects not conveyed by simple published specifications. The best way to ensure this is to order from the manufacturer or agent – this also prevents any infringement of safety standards; see Figure 20.7!

5

FIELD TIMEBASE

The field timebase in a TV, VDU/monitor, electronic viewfinder or tube-type camera is required to deflect the scanning spot in very linear fashion from top to bottom of the screen (target) area in a period of about 19ms. The other 1ms of the field period is occupied with the retrace or flyback period, during which the scanning spot is quickly returned to the top of the screen. All of the field timebase circuits of interest to us operate *magnetic* deflection systems, so the effective load of the field output stage is the field deflection coil – the yoke. The yoke appears to be resistive to the relatively slow-moving scanning stroke, but its inductance is significant during the twenty-times-faster flyback stroke.

Field timebase block diagram

The separate stages of a field timebase are less obvious than those of the line timebases described in the last chapter, especially in modern sets, where the entire field timebase is often embodied in a single small IC, with sync pulses entering on one pin and the deflection yoke hooked to another. Even so, all the components of the block diagram of Figure 5.1 are present, and need to be understood and identified for diagnostic purposes, even if this is confined to relating chip-peripheral components to the processes of timing, shaping, controlling and coupling the field-scan waveform. In older sets the separate blocks are easier to identify where discrete

circuits or a mixture of IC and transistor circuits are used.

Oscillator and sync

The process starts with an oscillator – *timing generator* is perhaps a better description – to initiate sweeps at 20ms intervals in accordance with the field sync pulses of the incoming video waveform. If the picture is rolling in either direction the frequency of the oscillator is incorrect. Normally the free-running frequency of the oscillator is arranged to be *lower* than the standard 50Hz, typically 43–48Hz. This means that each ramp is *longer* than 20ms until sync pulses are applied, and is then terminated early by the sync trigger.

This gives an important clue when analysing 'rolling picture' symptoms: a downward picture role indicates that the oscillator frequency has drifted high in frequency, suggesting a check of the oscillator's time-constant components, etc.; while a slow upward picture roll indicates a loss of sync. *Fast* upward roll is likely to be caused by the oscillator frequency drifting low, beyond the pull-in range of the sync pulses. Where a field hold control is provided, its adjustment can give useful clues – if the picture can be made to 'hover' or to roll in either direction with this pre-set, loss of sync pulses is indicated. In tracing this with an oscilloscope, it is easy to mistake spikes from the

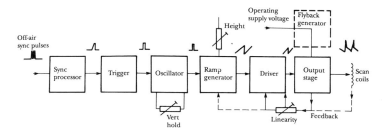

Figure 5.1 *Stages in the generation of a field scanning current*

field timebase itself for the sync pulses. When the picture is rolling any vestiges of oscillator pulses should travel horizontally along the scope trace, crossing the sync pulses themselves – or vice-versa, depending on which set of pulses the scope is locked to. Where no field hold control is provided the free-running frequency can be seen by removing the sync feed by some safe means.

Where the field sync is weak or non-existent, it is prudent to start by examining the quality of the video signal emerging from the vision detector – crushing of the signal anywhere upstream (i.e. i.f. stages, or due to a faulty a.g.c. system) can compress and distort the sync pulses to the point where vertical triggering is erratic, see Figure 5.2(a). The same results can come from poor l.f. response in the pre-sync-separator stages, leading to the waveform shown in Figure 5.2(b). Here the field sync pulse train is on a gradient and cannot be properly separated for integration: this can, for

example, arise from a 'dry' coupling or decoupling capacitor in video, i.f. a.g.c., or sync-separator sections.

Poor field sync, too, can result from incorrect bias on the sync separator input point, typically due to failure (H/R, high resistance) in resistor R702 in Figure 4.3. Many of these field sync problems, rather than causing continuous rolling, may give rise to vertical *judder*, where the whole picture vibrates vertically, often intermittently, and depending on picture content. A typical user-description is that the picture rolls on changes of shot, or judders on dark scenes.

Ramp generator

Returning to Figure 5.1, the output waveform of the oscillator is (in most designs) a short pulse every 20ms, and is used to close a flyback switch associated with a charging capacitor. The fly-back switch and charging capacitor together from a *ramp generator*. The ramp is formed across the capacitor by charging it from a constant-current source until the flyback switch across it closes, 'dumping' the charge in preparation for a new ramp. The charging current determines ramp amplitude, hence picture height; the shape of the ramp sets the vertical linearity; and upon the period of the ramp depends field frequency and hence vertical synchronization of the picture. These points are important when fault-tracing, since examination of the ramp with an oscilloscope will reveal whether the problem stems from the ramp generator section or some defect further downstream.

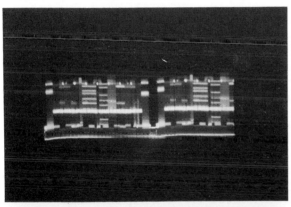
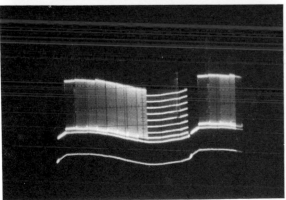

Figure 5.2 *Defective sync: at (a) the amplitude of the sync pulses is suppressed; picture/sync ratio should be 7:3. At (b) the d.c. level of the sync pulse train is wandering*

Driver stage

The ramp waveform is applied to a driver stage, which acts as a buffer between the high impedance of the charging circuit and the low input impedance of the field output stage. Almost invariably the driver stage is under the influence of feedback: a.c. for purposes of linearity correction and adjustment; and d.c. for stabilization of the operating conditions of the output stage. These feedback loops are crucial to the operation of the timebase, and should be checked early when the on-screen symptom is gross overscan (Figure 5.3) or very bad linearity. The driver stage is invariably d.c.-coupled to the output stage, so that a fault in one section upsets all voltage readings in the stage. For this reason, in the

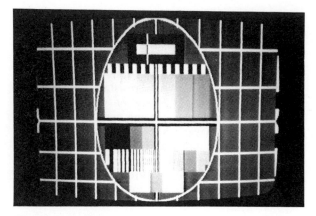

Figure 5.3 Gross vertical overscan due to insufficient negative feedback

event of a major fault in a discrete field output amplifier a quick approach to diagnosis is to unsolder all but one leg of each transistor or diode and check their junctions with an ohmmeter, replacing them as necessary.

Output section

In virtually all designs of field output amplifier the final stage is made of two transistors in series across the d.c. supply. The load (yoke) is capacitively coupled to their centre-point, and the output pair are driven in a way appropriate to their configuration – pnp/npn complimentary, two npn with antiphase drive, etc. As a general rule of thumb the *lower* of the transistor pair governs scan in the lower half of the screen, and vice versa. Thus top cramping may be due to a faulty diode or transistor in the upper half of the circuit, while poor linearity at the picture bottom is more likely to be due to a fault in the lower transistor. Bottom *cramping*, however, may well be due to poor regulation of the supply line to the output stage: the bottom of the picture generally corresponds to maximum current in the yoke, and thus to maximum demand on the voltage supply. Check with an oscilloscope for 'dents' in the supply rail voltage at field rate: the result, perhaps, of a faulty reservoir capacitor or similar problem in the supply circuit.

At about the centre of the screen, one of the output transistors hands over to the other, and if the transition is not smooth the effect is *crossover distortion* of the scanning waveform. It is caused by incorrect standing bias ('quiescent current') in the

output transistor pair and can usually be traced to the components associated with the bases of these two. The on-screen result is a compression of the scanning lines across the screen centre, often manifest as a thin dark horizontal line. The 'kink' in the scanning current waveform is surprisingly difficult to see on an oscilloscope display unless the centre portion of the output ramp is magnified by high X and Y gain settings.

When, as often happens, the yoke circuit becomes O/C (seldom does the yoke itself fail, the trouble is usually in a joint, plug/socket link or 'series' component) the output stage will continue to work, but with no load the output voltage swings rapidly between ground and supply-voltage potentials, giving the effect of a 50Hz square-wave on an oscilloscope display, whether the output stage consists of a discrete transistor circuit or IC device. The O/C point can be quickly traced with the oscilloscope; if the square-wave is present at both ends of the field scan yoke, the problem is in its ground path – typically an open or dry-jointed current sampling resistor. In the circuit of Figure 5.4 the yoke circuit consists of S10, scan coil L1, S12, L7A, L3, S11, scan coil L2, S9, coupler C29, the parallel combination of R59/60/61 and the soldered joints of all these components. The 'S' components are plug/socket connections. An O/C in any of these gives 'frame collapse' symptoms. In some sets the situation is revealed by a slight 'wiggle' of the white line at the extreme left of the screen – it's caused by a ringing effect in the field scan coils after line flyback.

Transistor field output stage

To illustrate some of the problems which can arise in the earlier field timebases, Figure 5.4 shows a 'discrete' design of driver and output stage. T7 is a voltage amplifier, d.c.-coupled to the whole of the output stage, with stabilizing d.c. feedback from the output stage via R52. Upon this resistor, and the network R47/48/49, depends the operating voltages throughout the amplifier, and in particular the mid-point voltage at T10 emitter. As well as d.c., a.c. voltage and current feedback are applied to T7 emitter, the latter via C23 from current sampling resistor R59/60/61. If one or two of these resistors goes O/C, the sawtooth sample across the now-higher resistance increases, and feedback action now considerably reduces the picture height as a result, though linearity remains good. Most sets

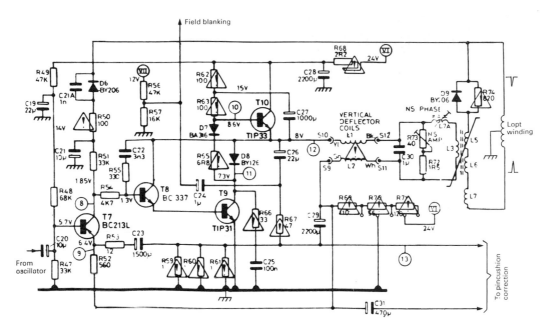

Figure 5.4   *Early form of field driver and output stage (ITT)*

using this type of sampling system have a single resistor in place of the three here. If it reduces in value, picture height increases and scan linearity becomes poor.

These older field timebase circuits make use of lots of electrolytic capacitors, which tend to dry up and lose capacitance as the years pass; certain types of capacitor are also prone to leakage. The results can vary from foldover and poor linearity to complete loss of field scan. A quick check for loss of capacitance is to bridge a substitute across the print connections of the suspect – the substitute does not have to be exactly the right value to prove the point.

Transistors are perhaps the next most vulnerable components, and their problems are most quickly diagnosed with a multimeter – on ohms range in 'cold checks' where the fault is a catastrophic one, i.e. R68 in Figure 5.4 overheating; and on volts range with power on where the fault is not destructive. Some examples: if all the voltages in the output stage are high, check T9 base voltage. If it's high, T9 b-e junction is open; if it's low (i.e. under 600mV) T8 is either O/C or not being turned on: check its base voltage, set in turn by T7 conduction. If, alternatively, the voltages in the output stage are low, T10 is probably not conducting: a high base voltage indicates a faulty

transistor, while low base voltage suggests R62 or R63 is O/C, or that T9 is S/C or turned hard on.

In general, S/C transistors in the field output stage cause excessive current consumption and invoke some protection artifice: here fusible resistor R68 would spring open. As with PSU circuits, innocent components are vulnerable to the effects of the crossfire, as it were, and a check of T7, T8, D7 and D8 is worth making when replacing faulty or damaged field output transistors.

Tracing field collapse

A very common symptom is lack of field scan, with just a thin white line across the centre of the screen. The line may not be visible at all: many TVs are designed to blank the screen altogether if vertical deflection fails, to protect the tube faceplate phosphor from burning. With these sets, advancing the A1 (screen) preset will generally make the line visible; artificially reducing one of the cathode voltages momentarily will have the same effect. Where no such protection circuits exists, the first priority is to protect the picture tube by reducing the brightness so that the line is just visible.

With a field circuit like that of Figure 5.4, first establish that the supply voltage (here VI, + 24V) is present, then measure the mid-point voltage at T10 emitter. If it's reasonably correct, there is either an O/C yoke circuit, or the field oscillator has stopped running. The oscilloscope should now be brought into play: in the one case it will display a squarewave at T10 emitter; in the other an absence of sawtooth drive at input coupler C20, leading to a check of supply voltage, transistors and passive components in the oscillator stage, in that order. If the mid-point voltage is incorrect, the d.c. feedback loop tries to restore correct voltage, which tends to exaggerate the incorrect voltage readings. If, for instance, T8 collector went O/C, the mid-point voltage would rise, and via R52 T7 would push up the base voltage of T8 in an attempt to restore correct conditions. Correct interpretation of this type of situation saves much time during diagnosis.

N/S raster correction

The older types of picture tube need NS (North–South) raster correction to cancel out pincushion distortion at the top and bottom of the picture. This is generally done by means of a transductor, represented in Figure 5.4 by L3/5/6. The latter two coils carry a line-rate waveform which becomes modulated on the field scanning current at the extremes of vertical deflection. Poor raster shape at top and bottom, where it cannot be corrected by amplitude and phase controls (here R73 and L7A respectively) is usually due to a faulty transductor (the early types especially) but check any series resistor in the line feed circuit, and any tuning capacitor associated with the phase control. The resistor across the field winding of the transductor (here R72/73) has a damping effect on the transductor; if it goes O/C, horizontal dark/light striations may appear across the top of the picture.

Later picture tubes have no need for NS raster correction.

Complementary field output amplifier

Before we leave discrete transistor field timebases, Figure 5.5 shows a once widely used form of class B field output stage. Virtually all the remarks already made apply equally to this arrangement. The output of TR1 is applied to the bases of both (complementary) output transistors. Feedback arrangements are similar to those of Figure 5.4, but

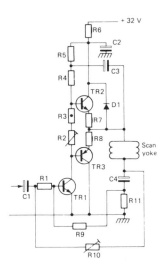

Figure 5.5 *Complementary-symmetrical field output stage*

the bootstrap capacitor C3 has particular relevance to the flyback, and so should be first suspect in cases of top foldover or superimposition of teletext pulses on the top of the picture (see Figure 5.9). R2 and temperature-compensator R3 are the ones to check for crossover distortion (see above); and decoupler C2 and yoke-coupler C4 for bottom cramping.

IC field timebases

The examples of field timebase systems given so far are now uncommon in new designs, having been ousted by various forms of IC-based deflection systems. In general, fault-tracing in these is simpler than for discrete types.

The energy required by some types of tubes was more than could be managed by an IC, leading to the adoption of the 'booster' type of circuit depicted in Figure 5.6. Here the negative-going ramp at pin 4 of the IC drives the yoke directly via R6 for small deflection currents (about the screen centre) and is assisted by current dumper transistors Tr1 (upper half of screen) and Tr2 (lower half of screen) for large deflection currents. Failure of R6 can cause cramping at screen centre.

Failure of components in the negative feedback circuits of output stages like that of Figure 5.6 can give rise to strange effects. The network C1/3/4, R4/5 is largely concerned with curtailing h.f. response and preventing instability or spurious oscillation, which may only be manifest as over-

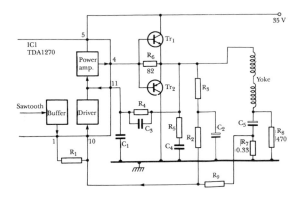

Figure 5.6 *Tr1 and Tr2 form a power booster to interface the field output chip IC1 to the deflection coils*

heating or sporadic failure of the IC or output transistors; depending on the severity, varying picture effects may be present, but the fault will always show on an oscilloscope. Individual components in these networks can occasionally develop leakage or 'noise' to cause picture jitter – aerosol freezer-spray generally pinpoints the culprit.

The main feedback path (bottom of diagram) carries d.c. and a.c. feedback to the IC, the latter developed across sampler R7 – see earlier remarks on negative feedback faults in the output loop. A current bleed for correct vertical picture-centring is provided by R8 – check it and coupler C5 in any case of a vertically-displaced picture.

A different type of IC/transistor field timebase design is shown in Figure 5.7. Here the IC contains both field and line oscillators as well as a field driver section and the sync separator. Starting again with sync, field roll or judder should direct attention to C601 and R603, perhaps the two most vulnerable components in the sync feed circuit. Much of the rest of the field timebase features are recognizable from previous diagrams and text, but this circuit has some unusual features worth bearing in mind during fault diagnosis. The field driver stage inside I501 is arranged so that a large fall in voltage at C510 shuts down the IC, reducing pin 2 voltage to zero. Thus if R513 goes open, for instance, drive ceases and Q502 cuts off to give puzzling readings on test instruments.

A protection circuit in this set operates by monitoring Q501 collector voltage. If it falls far, excessive current in the field output stage is indicated and a line oscillator shutdown section within I501 is brought into operation by the action of Q603. Going several stages further than beam-blanking, this shuts down the set altogether in the event of field collapse. For test purposes, the protection can be overridden (with caution) by grounding I501 pin 9, whereupon the fault will be revealed; in this particular design the protection may also be invoked by excessive picture-tube beam current or too high a level of line flyback pulses, monitored by ZD602 and D604 respectively. Forewarned is forearmed in cases like these, which only become obvious when the manufacturer's service literature is consulted. The most helpful manufacturers provide product-specific fault-finding hints and charts, and quoting specifically from that relating to Figure 5.7 we are advised, for low scan or poor linearity, to first try an adjustment of the height control. Possible results and likely culprits (applicable to other, similar circuits) are given as follows:

1 Top compression: check C510, R510 Q501
2 Bottom compression: check C514.
3 Low height: check R516, R517, R518.

These are probably based on practical experience with production sets, and should be regarded as supplementary to the general advice given throughout this chapter. The position of the height control across the yoke-current sampling resistor in this design, and the lack of any field linearity control, are unusual features very relevant to fault diagnosis.

TDA2600 IC

Sets using this switch-mode field timebase chip are now obsolescent, but may find their way onto the repair bench. Loss of field scan is usually due to failure of the chip itself or its connections to the PC board via a socket. Where the chip has failed, ensure that the replacement has a good thermal bond between its top surface and the heat-sink, and that its holder and joints are good. Before restoring power, check for poor contact in the riveted joints of the main reservoir capacitor in the set's PSU; and that the connections inside the *mains* plug are secure.

Single-chip system

A typical one-chip field timebase with peripheral components is shown in Figure 5.8. It is completely self-contained, having a sync input at pin 8, a

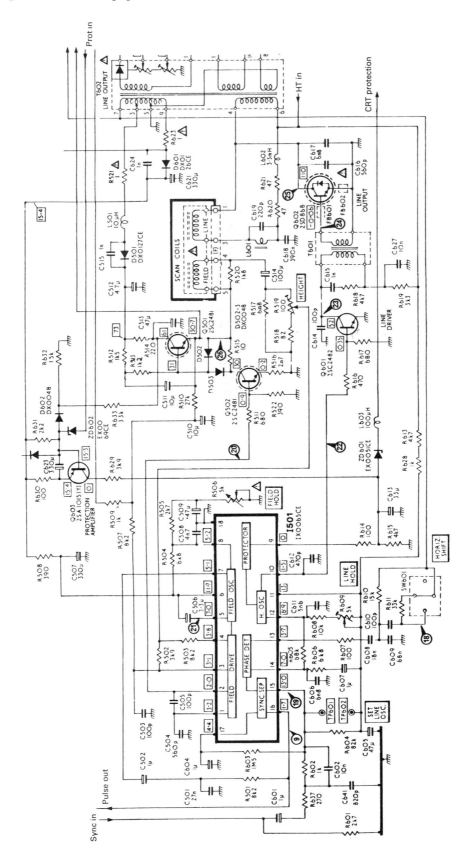

Figure 5.7 *Field timebase and associated components in the Rediffusion/Doric Mk5 TV chassis design*

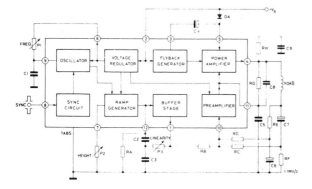

Figure 5.8 *Complete field timebase in one chip, also incorporating a 'flyback generator'*

power output point at pin 4, and operating voltage applied to pin 5 (output stage) and pin 2 (rest of chip). These supply voltages are the first to check in the event of low- or no field scan. The IC's internal circuit is grounded via its copper heat-spreading tabs.

Most of the other features of the circuit have already been covered in this chapter – this design has a minimum of peripheral components. P1 and C1 are the ones to check for incorrect field frequency; where no field sync is present, check for clean sync pulses at pin 8, but remember that they are derived and processed elsewhere, usually in the line sync/oscillator chip. Height problems, where the supply voltage to pin 2 is correct, should direct attention first to P2 at pin 7, then to the feedback circuit from RF (bottom RH corner of diagram), where the values of RB, RC and RF are critical. Linearity and jitter problems often arise from problems in the linearity correction circuit – P3 and C2/C3, which are sometimes electrolytic types. The anti-parasitic/h.f. roll-off components here are RG and C8.

The new feature in Figure 5.8 is the flyback generator associated with IC pins 3 and 5. C4 and DA together with the flyback generator within the chip operate to double the supply voltage to the output stage during the period of flyback. The action can be monitored with an oscilloscope at IC pin 5 where (with a typical supply voltage of +25V) pulses of about 1ms duration and 50V amplitude should be present; a similar pulse amplitude appears at the output point (here pin 4) during field flyback. The idea is to reduce power dissipation in the field output stage, and similar artifices (some using transistors) were incorporated in discrete field output circuits. Problems in this

section generally lead to defects at the *top* of the picture – top foldover, cramping or teletext interference *on picture* (Figure 5.9) should direct attention to the diode and capacitor represented here by C4 and DA. Short life of the field timebase IC can also be caused by faults in these devices – replace them both if in doubt. Top cramping, foldover and similar problems, whether or not a flyback booster circuit is incorporated, is generally due to slow or 'constricted' flyback, revealed by scope tests. In older circuits using inductive components in the field output stage, S/C turns can cause this problem.

Figure 5.9 *Teletext interference lines on picture: the dots and dashes continually sparkle and twinkle*

Two-chip field timebases

For maximum versatility in receiver design, current trends (at least for large-screen colour TVs) are towards separate field generator and amplifier ICs; this permits the use of line-pulse counter circuits to derive field timing directly for better noise immunity in the field flyback triggering. In these designs the 'field oscillator' becomes a digital appendage within a scan timing generator chip.

A field output IC's block diagram is shown in Fig. 5.10. All its peripheral circuits have already been described – the flyback booster R1/D1/C1 will be recognized from Figure 5.8; and the yoke circuit is now very familiar. Pin 1 is the key input point: here there should be a 3V peak-to-peak ramp waveform. From the servicing point of view the main points of interest are the protection systems built in. The SOAR (Safe Operational ARea) limiting section keeps the dissipation of the Darlington pairs which form the output stages within bounds at all times: this protects the device from

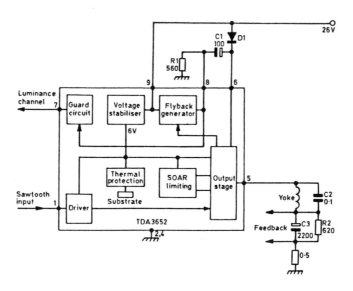

Figure 5.10 *TDA3652 purpose-designed field output chip with peripheral components*

S/C loads etc. The guard circuit is d.c.-coupled to the luminance section (usually another IC) and biases off the picture tube when no output current flows at yoke-connection pin 5. In some field IC designs the field flyback blanking pulses pass out of the guard pin too. This protects the tube from line-burn during field collapse.

Interlace

Poor interlace prevents the scanning lines of one field falling exactly in the gaps of the other, giving a coarse 'liney' appearance to the picture. It's caused by timing errors in flyback triggering, and may arise from a problem in the field sync pulse integrating circuit (R501/C501 in Figure 5.7) or from false triggering by stray line flyback pulses. The latter can be due to such diverse causes as poor lead dressing; inadequate grounding of PC boards;

or line pulses, coupled in the scanning yoke, entering the field oscillator via a feedback loop, a situation normally prevented by suppression capacitors wired across the field scan coils – C2 in Fig. 5.10 is an example. In modern TV sets with Teletext check that the receiver is not operating in 'de-interlace' mode due to a control line fault.

Blanking

The scanning spot must be fully extinguished while retrace is taking place to prevent spurious light-up of the screen with twinkling Teletext pulses, etc. The field blanking pulse is originated in the field timebase section and passed to the tube in a way depending on the vintage of the set – see the section on sandcastle pulse in Chapter 4, and the description of luminance paths in Chapter 6.

6

LUMINANCE AND DECODER

This chapter will concentrate on the processing of the picture signals from a basic CVBS (Chroma, Video, Blanking, Syncs) signal forward to the picture tube. It is applicable to TV receivers from the vision demodulator onwards; to TV monitors and viewfinders; and to the decoder sections of 'digital' TVs, VCRs and effects generators where separate Y, U and V signals are processed and stored. Most of these stages are relatively trouble free, working on small signals with low power levels. High voltages are necessarily used in tube-drive output stages to provide the required voltage swing for picture-tube beam modulation, and it is here, perhaps, that faults and breakdowns are more likely to occur.

The variation in circuit design in the signal-processing area of all the equipment likely to find its way to the repair bench is very wide, since much evolution has taken place over the years, culminating in the high-voltage RGB output chip capable of direct drive to the picture-tube cathodes. As in previous chapters the coverage of this one will be tailored to the most likely requirements of the 'GP' serviceman at the bench.

Block diagram

Figure 6.1 offers a block diagram of the stages involved in post-demodulator signal processing. All of them are present, though how obvious and visible they are depends entirely on the vintage of the set in question. Very early receivers use wholly discrete circuits in this area; the latest types have a single IC which carries out all the functions in Figure 6.1 apart from those of the high-level RGB output amplifiers.

The CVBS signal is first passed through a sharp 6MHz trap (in LC or, more usually, ceramic filter form) to remove the 6MHz sound carrier. Excessive sound carrier upsets tonal values of the brighter parts of the picture, and in monochrome and high-definition colour tubes gives rise to a fine *overall* dot pattern. The presence of sound carrier on the luminance signal is betrayed by a thickening of the horizontal lines (i.e. during the line sync pulse) of an oscilloscope trace; it can be minimized by tuning the intercarrier trap if it's adjustable, replacing any ceramic filter or associated components if not.

Next the luminance and chrominance signals are separated by filters. Following the luminance path, first is encountered a 4.43MHz notch filter to remove most of the colour subcarrier from the luminance signal. If it is mistuned downwards poor definition is the result – the higher luminance frequencies are attenuated. If the trap is mistuned upwards, or is not working, the chroma signal makes a dot pattern on the picture; its severity depends on how saturated the colour is, and it would be particularly noticeable, for instance, in the colour bars of the test-card or from a pattern generator. This dot pattern remains when the colour is turned down or off. To tune the notch filter, hook an oscilloscope to the luminance amplifier and adjust for minimum 'fuzz' on the stairsteps of a standard colour bar signal.

The luminance amplifier has several functions. Here is a gain-controlled amplifier for contrast setting: contrast is proportional to a d.c. voltage on the control line coming from the 'user-interface' section of the TV, and is generally also influenced by the beam current limiter (see Chapter 4). Also present is a clamp, wherein the signal corresponding to the black parts of the picture is restored to a fixed reference voltage once per line, during the back porch period. The reference potential is governed by the brightness control which thus moves the entire luminance signal up and down in voltage – from here forward to the picture tube the signal is d.c.-coupled. Again the brightness control line is a d.c. one with brightness proportional to applied voltage. Incorrect brightness level can stem from many causes, but a common one is wrong standing voltage at the tube cathodes due to a problem in the controlled clamp. Check the bright-

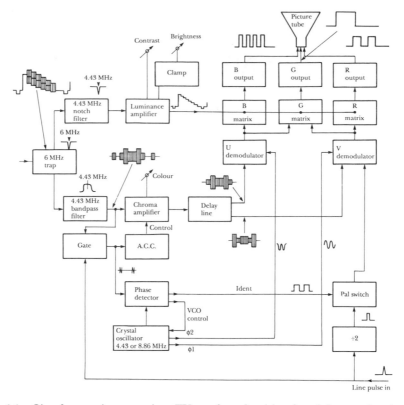

Figure 6.1   *Signal processing stages in a TV set, from the vision demodulator to the picture tube*

ness control voltage against that quoted in the manual, and also the clamp reservoir capacitor which holds the black level reference voltage – this is often a tantalum or aluminium electrolytic type. We shall look at specific circuits later in this chapter.

The output of the luminance amplifier in a monochrome outfit feeds a voltage amplifier whose output is applied to the picture tube cathode. In a colour set a low impedance feed is provided for the *matrix* where the Y signal is added to the B-Y, G-Y and R-Y signals to form RGB outputs for amplification to the level required by the tube cathodes.

In a colour set there is a short delay line (about 700ns) to bring the wideband luminance signal into time-coincidence with the chroma signals, whose passage is slowed by their narrowband path. An open-circuit delay line will delete the luminance signal (though 'edges' may get through) to give a vague 'transparent' picture consisting of large areas of ill-defined colour. If (as is often the case with IC designs) the d.c. level of the luminance signal is

established *before* the delay line, a break therein will result in no picture of any kind – the screen will not light up. Occasionally, due to a dry-joint perhaps, the delay line will come 'off-earth' to give a ringing symptom like that shown in Figure 6.2.

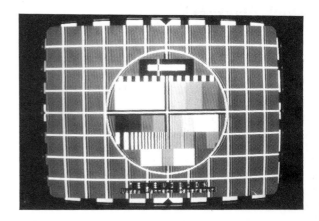

Figure 6.2   *'Ringing' effect on picture*

The chroma path starts with a bandpass filter centred on 4.43MHz. At its output all traces of luminance and sync should be absent to give a waveform on colour bars similar to that shown at the appropriate point in Figure 6.1. An oscilloscope picture is given in Figure 16.8. The pure chroma signal now goes in two directions: to the reference chain and to the signal amplifier. The burst gate needs line-rate pulses to key out the bursts; they come from the line oscillator or output stages, and if they disappear or are mistimed there will be no colour.

Every colour decoder has a crystal, resonant at 4.43MHz or 8.86MHz to generate a local reference c.w. signal. If the crystal oscillator stops, colour will disappear, and the same applies (in most designs) if it goes off frequency – once the loop comes out of lock the colour killer comes into operation. In some decoders it is possible to override the colour killer (see service manual) to observe the floating, unlocked colours which result from an unsynchronized subcarrier oscillator. If the correct colour can be made to 'run-through' by adjustment of the frequency pre-set trimmer the problem is in the burst gate or phase-detector sections; if the crystal frequency, by observation of the TV screen (killer overridden) or by the reading on a counter, is not correct, the culprit is more likely to be the crystal itself, though an attempt at adjustment of its frequency by the adjacent trimmer should first be made.

The correctly phased crystal oscillator provides two c.w. drives, 90° apart, to drive the demodulators. In early designs this quadrature condition was set up by manual adjustment of a coil, typically adjusted for maximum output from the demodulator it drives, or for specific conditions on a special test pattern – see decoder alignment instructions or those of the pattern-generator manufacturer. Where an 8.86MHz crystal is used the correct phases of subcarrier are produced without the need for critically selected or adjusted analogue components.

The second route for the chroma signal is via the chrominance amplifier, which is voltage controlled, and whose gain is under two influences, those of the user saturation (colour) control, and the a.c.c. (automatic colour control) which regulates the chroma signal according to the reference provided by the amplitude of the burst signal. The effect of the a.c.c. regulation is that as the chroma signal becomes weak there will be no visible desaturation: the colours will not become paler. What happens is that the colours become progressively noisier and

more confetti-strewn until they suddenly disappear altogether as the colour killer comes into operation. As we shall see, some IC designs pass the burst signal through the chroma amplifier and the delay line.

The delay line itself feeds a matrix from which two outputs emerge: U and V, to become B–Y and R–Y respectively after de-weighting. Lack of one of these colour-difference signals (examine a colour bar display) could be due to a demodulator problem, but first examine the subcarrier signals emerging from the delay line matrix, and the local subcarrier feeds to the demodulators – the latter are accessible in discrete and three-chip decoders.

Problems in the delay-line itself and its matrix lead to Hanover bars on the display, best seen in a plain red raster or on colour bars: the effect is changes of hue over alternate horizontal line pairs. Adjustment of the delay line matrix (R and L elements) may clear this (follow maker's decoder alignment instructions); if not, the delay line itself is probably damaged or faulty. Check all print and connections before condemning it.

If the PAL switch (bottom right of Figure 6.1) is not working the effect is similar to Hanover bars, but mainly shows on red colours as alternate lines of green. A PAL switch working in the wrong phase completely reverses the colour display, interchanging green and red in effect. Symptoms like this should direct attention to the line pulse feed to the switch and, where it is accessible, the ident stage. Such problems are rare in IC decoders, but can arise through failure of external components (capacitors etc.) used for coupling, decoupling and pulse feed. Most manufacturers provide a block diagram of the individual stages within their ICs, and this is a useful guide to the functions of the components hooked to the chip pins – examples of these are given later in this chapter.

The RGB output stages operate from a high-voltage line and provide peak-to-peak output voltages of about 70V for the three cathodes of the picture tubes. They are closely matched in respect of gain, frequency response and operating voltages: if their gains differ, grey-scale tracking errors arise to upset black-and-white and colour reproduction, especially in picture highlights; if their operating voltage levels differ, colour tinting in the low-light areas results; and if their frequency responses are not identical fine detail areas of the picture have colour tints or smears. A very restricted frequency response in one of the three amplifiers smears the affected colour to give a display like that arising from misconvergence, but confined to *vertical* lines.

A switched-off output stage raises the voltage on its tube cathode to cut off the beam and delete that colour altogether. A turned-on or short-circuit output stage reduces tube cathode voltage to turn the affected gun of the tube hard on and suffuse the screen with light of that colour – beam-limiting or overcurrent protection (via the EHT supply) may well delete the other colours or inhibit PSU operation, an action betrayed by the time-delay until the tube has warmed up. Further possibilities for display and raster problems are discussed in Chapter 10.

Such discrete transistor designs as still exist follow the block diagram of Figure 6.1 fairly closely. In IC-type decoders the individual functions are less obvious, and may be distributed amongst one, two or three chips. Some representative examples, spanning several years, will be given next.

Three-chip decoder

The first all-IC decoders used three chips in a configuration like that of Figure 6.3. The first device, TBA560C (IC1) handles most of the luminance processing, so is primary suspect for brightness and contrast faults where conditions on the tube cathodes are incorrect. IC-external suspects for incorrect black level are the capacitor at pin 4 (open-circuit gives bright raster) and the brightness control potential at pin 6, which should be around 1.4V for normal operation. Incorrect contrast may well be due to wrong voltage at pin 2 (should be 3.4V approx.) which is partially derived from the beam limiter circuit; normal voltage here may mean that the video input signal is wildly incorrect due to a problem in the i.f. or a.g.c. sections – check for 3V p-p input at the emitter follower on the left of the diagram then if necessary consult the next chapter. Clean luminance signals at about 1.5V amplitude should be traceable from pin 5 through the delay line and into pin 1 of the TCA800 chip IC3. If not, another possible fault source is the blanking signal applied to pin 8 of IC1 which must be in accordance, voltage-level wise, with the manufacturer's waveform specification: this can only be checked with an accurate d.c.-coupled oscilloscope.

Turning now to the colour circuits, the filtered chroma signal enters on IC1 pins 1 and 15 for a.c.c. control (from a burst level detector in IC2, TBA540) via pin 14, which is a useful check-point in cases of

no colour. It normally sits at 1.1V, but rises to about 4V when the burst signal is absent or the TBA540 IC2 is faulty. With correct a.c.c. conditions a gated burst output should be traceable on pin 7 of IC1 and into pin 5 of IC2. Saturation control is carried out by varying the voltage applied to pin 13 of IC1: the higher the voltage here the greater the saturation, 3.6V being a normal level. The colour-killer operates on this pin, taking it low to delete colour. For checking, the killer can be overridden by applying 3 or 4V to IC1 pin 13. If this brings up unlocked colour either the crystal at IC2 is running at the wrong frequency, or the burst sample is defective, perhaps as a result of faulty components between IC1 pin 7 and IC2 pin 5. There should be over 1.5V of gated burst on this path. If, however, overriding the colour does not restore any form of colour, it is necessary to resort to voltmeter and scope checks of conditions on the IC pins with reference to the internal blocks of each chip.

Weak colour locking can be due to faulty filter components between pins 13 and 14 of IC2: in practice, this usually means intermittent colour, with more revealed once the killer is overridden.

The third chip, the TBA800, is devoted to demodulation and matrixing, and can be responsible for faults on *individual* colours, fairly easily traced by voltmeter and oscilloscope checks. Not shown in Figure 6.3, but very relevant to fault diagnosis, are the reservoir capacitors associated with the R, G, and B clamps at the right of IC3. These small electrolytic capacitors hang on pins 2, 4 and 6, and are worth checking when one colour is incorrect.

Video output stage

One of the simplest forms of video output stage is illustrated in Figure 6.4. This class A type is used in budget TV sets – three identical amplifiers for colour sets of course. If the collector voltage of T505 is low, feel R569 with your finger. If it's cold (and so long as it has 225V at one end and low voltage at the other) the resistor is O/C. If, alternatively, it is very hot T505 is passing excessive current. Short its b-e leadouts: if collector voltage now rises, the transistor is being driven too hard (T504 O/C or off earth?) or its emitter-to-ground path is too short. Check C553, R571 and the 'set black level' network R14/R16. If the collector voltage does not rise with T505 b-e shorted, the transistor is leaky or being bypassed elsewhere, i.e.

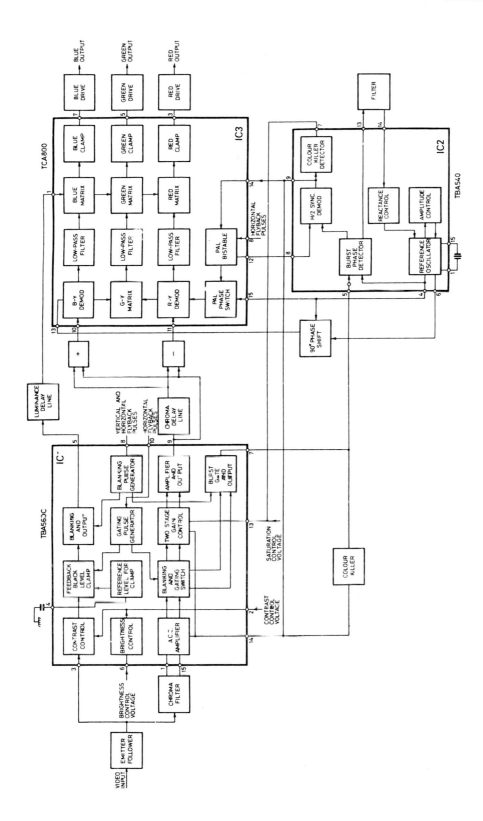

Figure 6.3 *Block diagram of an early colour decoder using three chips (ITT)*

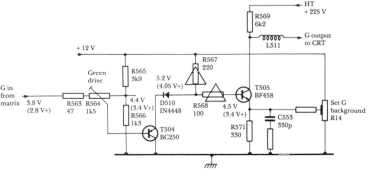

Figure 6.4 *Class A video output stage of a type widely used in colour and monochrome TV sets (ITT)*

a leaky spark-gap or cathode leakage/short at the picture tube.

If C553 goes O/C the frequency response of the stage is severely curtailed, leading to the problems described above: 'tinted' detail, smearing and misconvergence effect. Two possibilities in this circuit for complete lack of output are an O/C wiper at the drive control R564, and an O/C tube feed choke L511: if the cathode of a picture tube is allowed to 'float' the gun turns off and ceases to emit. If the *grid* of the tube 'floats' the gun turns on and floods the screen with light.

Two-chip decoder

Figure 6.5 shows a decoder using two ICs. It is similar in some respects to that of Figure 6.3, and many of the same remarks apply, though the functions of IC2 and IC3 in that figure are here incorporated into a single chip TDA2523, IC223. Unlike the previous design, this one passes the burst signal through the chrominance amplifier and delay line, though gating is employed to ensure that the burst level remains constant while the chroma signal in the same path (IC192 pin 6) is varied by

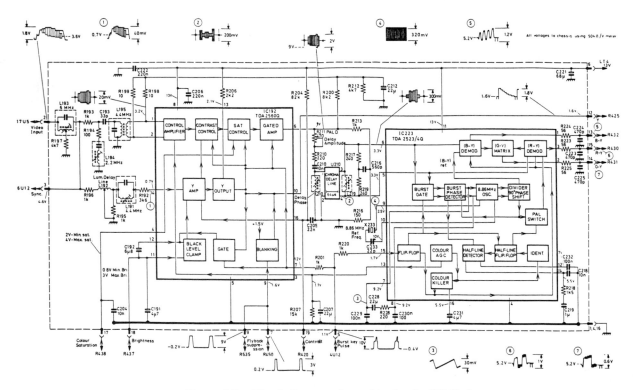

Figure 6.5 *Circuit diagram of two-chip decoder (Philips)*

the saturation control potential on pin 4 of the chip. This gives a useful clue in the case of 'no colour' symptom: if burst is present at the delay line, it's likely that control pin 4 voltage is low, due perhaps to C204 S/C or some external cause around the saturation control.

Sandcastle pulse

Many of the functions inside IC192 are controlled by the *sandcastle pulse* applied to pin 5 of the module. This rises from the 400mV level and has two components: a broad one corresponding to the line blanking period and having an amplitude of 3V; and a short one, timed to coincide with the colour burst on the back-porch, which momentarily boosts the pulse level to 10V. This composite pulse is widely distributed within the two decoder ICs as Figure 6.5 shows, and its integrity should be the first scope check for problems like no colour, incorrect clamping or blanking, etc. Here the capacitors associated with clamping and brightness level are C192 and C191 – check them for shading, brightness and intermittent flashing problems.

Reference chain and matrix

The TDA2523 chip IC223 processes the signal from delay-line matrix onwards as Figure 6.5 shows. The colour killer operates on the U and V demodulators, and depends on the presence or absence of burst as measured by the voltage at pin 14. The colour killer here can be overridden by grounding pin 16 of the chip.

In this and some other designs the primary-colour (RGB) matrix is outside the IC. The Y output leaving pin 12 of the module in Figure 6.5 is applied to an emitter-follower whose low output impedance drives a resistor network from which RGB outputs are produced. An alternative arrangement, used in other two-chip and many discrete designs, is to apply the Y signal (usually through separate drive control presets) to the emitters of the separate RGB amplifiers, and the individual colour-difference (R-Y, G-Y, B-Y) signals to their bases. Whenever the emitters of the RGB output transistors return to ground via a common circuit or component (typically a 7.5V zener in some older circuits) this should be checked for any 'brightness' fault where the tube cathode voltages are all incorrect.

RGB output stages

The simple output stage shown in Figure 6.4 has many variants. One type of class A output arrangement has d.c. feedback into the matrix IC to maintain correct potentials on the tube cathodes – the arrangement was amongst the earliest attempts to stabilize black level. Where a thick-film collector load resistor array is used this is the first suspect for faults involving an imbalance between R, G and B channels – closely followed by the associated fixed resistors. While the problem can be diagnosed by analysing voltages, it is often quicker to check the afflicted channel(s) by visual and ohmmeter inspection of individual resistors or semiconductors.

To give an idea of the way in which the screen display is affected by various circumstances of faults and missing colour components, Figure 6.6 shows what happens to each of the standard colour bars under the conditions specified on the left. The chart does not, of course, take into account any changes of d.c. level coinciding with faults in B-Y or R-Y channels.

A very quick check for the location of a fault in one of the RGB channels is to interchange signal feeds – to the tube, to eliminate faults in the latter; and to the output stages, to differentiate between faults in the drive (IC) and output sections. To take a particular example, imagine that the screen is very strong on red. Swap red and green tube cathode feeds. If the screen remains red, the fault is in the immediate area of the tube – a base-panel component, an incorrect voltage on grid or A1 (where they are in separate form) or a faulty tube. If the screen turns green, however, the problem is with the drive section, and if necessary interchanging G and R feeds *to* the output stages will further narrow down the field of search. These tricks are particularly useful where the fault is an intermittent one, but only work where (a) no feedback circuits are employed; or (b) the effect of any feedback circuit is taken into account. When transposing drives in this way it is important to analyse carefully the effect, perhaps in conjunction with the information in Figure 6.6.

The substitution practice is also useful in RGB amplifiers using more complex RGB output stages with class A or class AB multi-transistor designs. Since all three are identical, components (especially semiconductors) can be swapped between channels to isolate the faulty one. This is particularly relevant where an auto grey-scale-shift circuit is in

Colour bar hue	White	Yellow	Cyan	Green	Magenta	Red	Blue	Black
Tube guns on for correct hues	R,G,B	R,G	B,G	G	R,B	R	B	None
Main colour-difference signal(s) present	None	-(B-Y)	-(R-Y)	-(B-Y) -(R-Y)	B-Y R-Y	R-Y	B-Y	None
Blue gun only on	Blue	Black	Blue	Black	Blue	Black	Blue	Black
Red gun only on	Red	Red	Black	Black	Red	Red	Black	Black
Green gun only on	Green	Green	Green	Green	Black	Black	Black	Black
Blue gun off	Yellow	Yellow	Green	Green	Red	Red	Black	Black
Red gun off	Cyan	Green	Cyan	Green	Blue	Black	Blue	Black
Green gun off	Magenta	Red	Blue	Black	Magenta	Red	Blue	Black
No B-Y signal	White	Off-white	Green	Green	Red	Red	Black	Black
No R-Y signal	White	Yellow	Pale mauve	Dark green	Blue	Black	Blue	Black
R-Y signal out of step	White	Off-white	Magenta	Dark orange	Blue	Dark green	Purple	Black

Figure 6.6   *The effects on displayed hues of various decoder faults, using a standard colour-bar signal*

operation (details later), or where the requisite high-voltage transistor types are not in stock at the workshop!

One-chip decoder

Several variants of one-chip colour decoder have emerged in recent years. The first of them had functions similar to the decoders already discussed, with 'straight' RGB outputs to the tube-drive stages. These are similar, for fault-finding purposes, to the multi-chip decoders already discusssed. Where separate RGB inputs are provided for text injection there is a data blanking input to the chip. Its presence and effect should be borne in mind when tracing strange no-picture or partial-blanking faults. In general, unplugging the interface or text panel will eliminate problems here. A great simplification of the signal processing section of a

TV or monitor is conferred by the one-chip decoder. Servicing, too, is generally simpler, though it is important to check and analyse IC pin voltages and waveforms – the fault is not *always* in the chip, whose cost is high enough to make a wrong diagnosis embarrassing!

Second-generation one-chip decoders have an automatic grey-scale correction feature to take up tolerances and ageing effects in the picture tube. A popular example is the Mullard TDA3562A, a block diagram of which is given in Figure 6.7. The left-hand side of the IC diagram is conventional in modern IC practice, and we shall confine ourselves to the processing downstream of the RGB matrix. The video switching stage contains three changeover switches, governed by the voltage at pin 9: when it is low, internally decoded signals are passed out of the chip, and when it is high the chip is ready to accept RGB signals from a text decoder or external device (i.e. computer) at pins 12, 14 and 16.

In order to be able to control brightness and contrast of these and text signals, the contrast and brightness control lines (IC pins 6 and 11 respectively) operate separately on the three RGB channels. Next comes the automatic grey-scale control section. Short pulses are output on TV lines 23 (red), 24 (blue) and 25 (green), all in the field blanking period. These can be observed on an oscilloscope triggered at field rate, or at the top of the screen if picture height is reduced. Each cathode's current is measured during its short test pulse, the result passing back into pin 18 of the IC. This sample potential is compared with a reference voltage of 3.48V (generated within the IC and held as a charge on the capacitor at pin 19). The results are stored for each of RGB on IC pins 10, 20 and 21 respectively, so that shading, absence or mis-

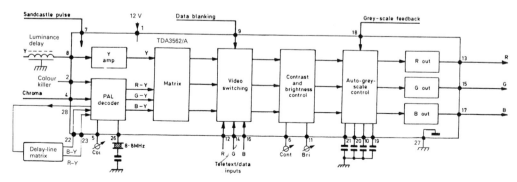

Figure 6.7   *Block diagram of the TDA3562 colour-decoder chip, with emphasis on the output stages and auto-grey-scale correction*

behaviour of one colour only should direct attention to the appropriate one. So long as the picture grey-scale is correct, a difference of greater than 200mV between these three pins indicates that one or more tube guns is low in emission.

The main problem from the servicing point of view is that almost any fault from the picture-tube backwards has the effect of shutting down the IC to give a no-picture symptom, though a very dim raster may be discernible in a blacked-out room. First check that the luminance signal is entering the chip at pin 8 and that pin 9 is low. These points will usually be correct. Next measure the 12V supply at IC pin 1: it must be between 11.5 and 12.5V to ensure correct operation, and failure of a 12V regulator can push it outside these limits. The next step is to carefully check the sandcastle pulse applicd to IC pin 7. This is shown in Figure 6.8(a), where + 2.5V is present during the field blanking period; if the + 2.5V level is continuously present the likelihood is that the blanked screen is masking a field-collapse situation – see Chapter 5. The line-rate display of the same sandcastle waveform at the same point is shown in Figure 6.8(b). For the duration of line blanking the voltage level rises to +4.6V, and for the short period of the colour burst the pulse level peaks at 11V, used within the device as a reference point for black level clamping as well as for colour burst blanking. Unless the timings and levels of these sandcastle components are correct the chip mutes all output signals.

The other 'regular' cause of IC shutdown is incorrect potential on the auto-tracking feedback pin (18) of the IC, and this is often caused by faults in the RGB output stages. A simple way of overcoming this problem is to override the mute action by applying 5V to IC pin 18 from an external source. Thus fooled, the IC should drive the output stages to give a raster and permit further checks. If no raster appears the problem is not in the auto-GST section. If required a more elaborate 'dummy' RGB output stage can be made as shown in Figure 6.9. This satisfies the IC's need for feedback signals and permits oscilloscope examination of the RGB, blanking and auto-GST pulse waveforms.

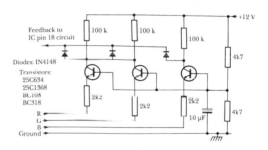

Figure 6.9 *Dummy RGB output stage to keep the auto-GST section working during faultfinding*

The diagram of the IC in Figure 6.7 has been drawn with some 'key' peripheral components to assist in fault-finding. This decoder design, like the two-chip type described earlier, routes the burst signal at constant amplitude through the chrominance amplifier and glass delay line. Bear in mind, too, that incorrect voltage at pin 25 will put the chip into NTSC mode and thus lose colour from PAL signals.

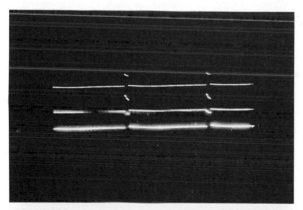

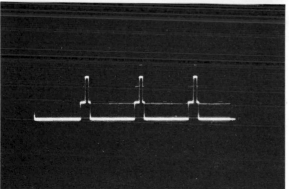

Figure 6.8 *Three-level sandcastle pulse:* (a) *field rate,* (b) *line rate*

Miscellaneous faults in Y/decoder circuits

Some common faults in the areas covered by this chapter and quick checks to trace them are as follows.

No picture

Check any video mute circuit and whether the picture signals are routed through a teletext decoder or its interface, which may be faulty; check whether the set is wrongly switched to AV operation, either from the control panel/interface or via an AV input socket status-switching voltage. Details of two common AV socket pinning systems are given in Figure 6.10. Check tube electrode voltages – see Chapter 10.

Flyback lines on screen (see Figure 10.8)

Check field blanking, but bear in mind that

excessive A1 voltages on the picture tube is a common cause of this.

Problems in green channel

Some sets have on-screen character generators with access to the green (sometimes other or all!) output channel. Problems here can affect the whole screen area.

Poor definition (Figure 6.11)

If no colour is present the fault will be in the pre-detector stages (see Chapter 7). Where colour is present and blurring is confined to *vertical* lines, use scope and multiburst signal for checking. For poor definition of vertical *and* horizontal lines, see Chapter 10.

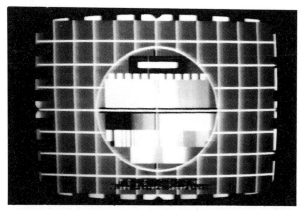

Figure 6.11 *Impaired horizontal resolution due to insufficient bandwidth in the video signal path*

General

The best tools for faultfinding are a scope and colour bar signal once it's established that the fault is not an obvious one. Check (a) supply voltages; (b) luminance and chrominance input signals; (c) pulse inputs; (d) IC pin voltages and waveforms in conjunction with the setmaker's block diagrams and pin data. In cases of difficulty it pays to read the circuit description which most (!) manufacturers provide.

Amongst peripheral components, the most fault-prone are small ceramic capacitors, electrolytics of all types, and print (physical) defects.

Adjustment of grey-scale tracking pre-set controls will be dealt with in Chapter 10.

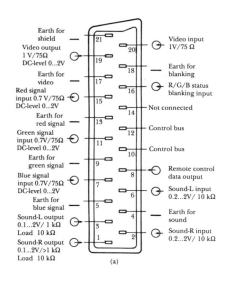

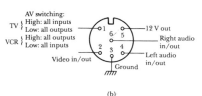

Figure 6.10 *AV interfaces:* (a) *SCART:* (b) *6-pin DIN*

7

TUNERS AND VISION RECEIVERS

The pre-demodulator signal processing stages of a TV set or VCR deal with small signals, dissipate very little power, and so lead a quiet life, making for good reliability. When problems do occur they may well be external to the actual signal-handling stages – in the aerial system, in the tuning voltage source or in a pulse feed to a keying or gating system. This chapter will examine, from a fault-finding point of view, the receiving system from the aerial to the vision demodulator.

Block diagram

Figure 7.1 shows the main processes covered in the descriptions which follow. The aerial signal is applied to the tuner, wherein it is selectively amplified then applied to a mixer, whose other input is a locally generated c.w. signal. The signal selected as output is the *beat product*, the difference between the incoming carrier wave and the local oscillator frequency. This intermediate frequency, *i.f.* is usually 39.5MHz in the UK, 38.9MHz in sets designed for use in continental Europe and else-where. By varying the tuner's oscillator frequency, any required channel can be converted to this frequency to be dealt with by a fixed-tuned i.f. amplifier. Oscillator frequency is governed by the user's tuning control.

The tuner's output contains various frequencies, wanted and unwanted: those required (vision and sound carriers) are selected and amplified, and those not required (adjacent channel carriers) are rejected. The i.f. filtering can do nothing for *image* frequencies, corresponding to a carrier frequency ten channels above the wanted one; and ± 4 channel sound carriers, any of which can cause beat patterns on the picture if not sufficiently rejected by the *tuner*. The two wanted carriers, a.m. vision and f.m. sound, pass together through the i.f. amplifier, and are regulated to a constant level by an a.g.c. system before being demodulated to produce a baseband vision signal and an f.m. intercarrier sound signal.

Aerial

Since the proliferation of terrestrial transmitters (in the UK, at any rate) the aerial and its installation is much simpler than in early days. Aerial faults give rise to three main symptoms:

1 Snowy pictures due to low signal strength: most often caused by corrosion of the connections at the aerial, deterioration or damage to the down-lead, or the aerial having moved off-beam. Before condemning these, check the connection

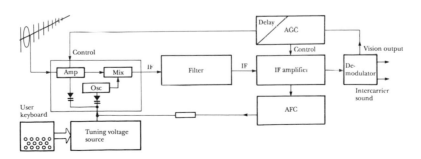

Figure 7.1 *Receiver section of TV or VCR*

to the plug at the bottom end of the cable, and the condition of, and supply voltage to, any mast-head or distribution amplifier.

A signal strength of 1mV should ensure a noise-free picture, but in practice it is quicker (and more convincing for the customer) to hook up another (portable?) TV set to check conditions.

2 'Ghosting' due to multipath reception as a result of the aerial moving off beam, the use of indoor or loft aerials, or the proximity of tall buildings, trees and natural features. The broadcasting authorities can offer much advice on this subject: addresses in Appendix 3.

3 'Garbledegook' on teletext reception – see Figure 11.3. Very short-term ghosting effects can have the effect of overlaying adjacent text data pulses to give erratic text reproduction. Where the cause is outside the TV set, the problem is more likely to be in the downlead or distribution system (impedance mismatch, poor lead routing) than the aerial element itself.

Tuner

Though the circuitry controlling it has changed much, the varicap tuner itself has in principle remained much the same for years. Using the circuit of Figure 7.2 as an example, and taking up the 'snowy picture' symptom from the previous section, the tuner could well be responsible, but before changing it, check that:

1 The aerial socket and its connections are in order. *Ensure* that the chassis is isolated, then connect the aerial lead directly to the tuner input. This also checks the separate 'boosters' used in the r.f. circuits of VCRs.

2 The tuner gain is being turned down by a faulty a.g.c. control line. Where information on the correct a.g.c. pin voltage is not immediately obvious, the quickest check is to apply a variable voltage from an external source, in this case to pin 2 of the tuner. If the picture does not

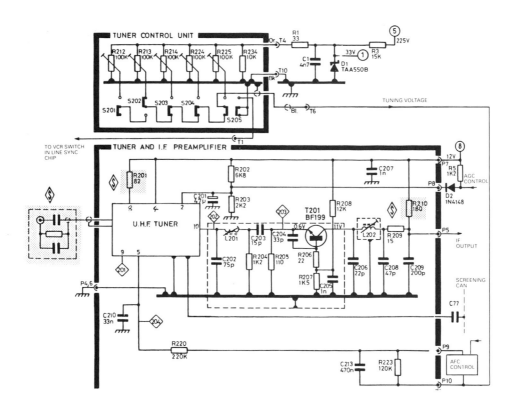

Figure 7.2 *UHF tuner and peripheral components. The separate i.f. preamplifier circuit around T201 is unusual (ITT)*

improve at any applied voltage between say 1 and 6, the tuner is suspect.

No signals

In a no-signal situation the tuner is implicated only if the screen is suffused with snow: a loss of signal further downstream does not permit much 'noise' to reach the screen. First make the aerial and a.g.c. checks detailed above. Check also that the tuner has correct operating voltage (here 12V at pins 4 and 8) and then check the tuning voltage at pin 5. If it is absent or restricted the tuner is looking at an empty channel. Check that 33V is available via R3 at stabilizer IC D1, then trace it through the pot-bank and channel selector switches, through the AFC section and R220. If the available voltage is all dropped across R220 it's likely that C210 or pin 5 of the tuner is S/C. If, alternatively, the tuning voltage pin is stuck high, the bottom of the pot bank (point T10) has come off ground. With all tuner pin voltages correct, the complete absence of signals suggests a faulty tuner, though in the particular case of Figure 7.2 it's worth checking the electrode voltages of RF pre-amp T201. with multi-band tuners make sure that the band-select inputs (switching pins) are correctly energized.

Tuner repairs are seldom justified these days – replacement is often cheaper and certainly more reliable, especially for the more common types, which can be bought cheaply from component distributors and wholesalers. Before doing so it's worth taking the covers off a suspect tuner and making a visual check for obvious damage or bad joints.

Tuning drift

The diagram of Figure 7.2 represents the simplest possible method of deriving tuning voltage, and is much used in early TV and VCR designs, and in later 'basic' models, in conjunction with an a.f.c. voltage derived from a resonant circuit near the vision detector. A common problem with this type of circuit is tuning drift. The first diagnostic action is to establish whether the tuner or the tuning voltage source is responsible. Monitor the tuning voltage (varicap control pin on tuner, here no. 5) with a digital voltmeter. If the tuning voltage decreases as the set drifts up the band (into heavy coloured patterning) the tuner is responsible. If the

tuning voltage increases as the set drifts down the band (into monochrome) the tuner is implicated likewise. If, alternatively, the tuning voltage *follows* the direction of tuning, the fault may be electrical leakage in the tuner itself – a microammeter in series with the tuning voltage input pin shows an increase in current under these circumstances, condemning the tuner.

Where the tuning-line current does not rise, but the tuning voltage is varying, first check (DVM again) that the stabilized supply voltage (D1 cathode in Figure 7.2) is constant, then suspect the tuning pot-bank and switches. A temporarily-wired 20KΩ pot in place of the pot-bank/switch system will prove the point, and can also be used for a tuner-exoneration test where more sophisticated tuning voltage sources (see below) are used. Always check for tuning drift with the a.f.c. off (switch, control cover-flap, etc.) to avoid disguising the fault, but expect a small degree of drift as normal with no a.f.c. action, and bear in mind that a drifting a.f.c. circuit (faulty diode, leaky capacitor?) will pull the tuning away from the correct point. If the tuning is pulled off the correct point each time a.f.c. is reinstated, the a.f.c. resonant coil (L36 in Figure 7.7) needs adjusting for spot-on tuning.

Electronic channel selection

Many tuning control systems use semiconductor switches in the channel selection circuit. A representative diagram of this type of arrangement is given in Figure 7.3, where the top ends of all the tuning pots are commoned to the + 33V supply, and the bottom end of the required pot is grounded as required by a transistor, usually within an IC as shown here. Problems in this type of circuit can arise as follows:

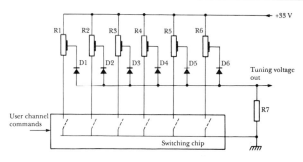

Figure 7.3 *One form of electronic switching for programme selection. The unused diodes remain reverse-biased*

1 Entire circuit at 33V, no signals: bottom IC lacking reset or command pulse (typically from another chip) or faulty.
2 Tuning drift, inability to tune: one of diodes D1 to D6 leaky, allowing an 'open' pot to influence the tuning line. The culprit is discovered by seeing which (unselected) pot has an effect on tuning line voltage; by individually disconnecting and checking the diodes; or in intermittent or obscure cases, changing all of them at a cost of a few pence each!

Sometimes the circuit of Figure 7.3 is arranged 'upside-down' with the electronic switches between the + 33V source and the tops of the pots. All the same comments apply, and in either type check that R7 or equivalent is not O/C when tuning drift is the problem: this resistor is essential to keep the selected diode turned fully on.

Some early forms of touch tuning relied on finger conduction between metallic pads, or hum injection by touch. Common problems were sticking or jumping to one channel caused by dirty deposits on touch pads, or faulty neons where used; and erratic operation due to failure of high-value (MΩ) resistors.

PLL tuning systems

Some later types of TV and VCR use frequency-synthesis tuning systems in which individual tuning pots for each channel are not required – the tuning data for each programme selection is held in a solid-state memory, and the tuner's local oscillator is 'steered' to the required frequency by virtue of its inclusion in a phase-locked-loop, PLL. A block diagram of the system is given in Figure 7.4, which is representative of several types and vintages of this tuning system. It is not practical here to describe its operation, covered in manufacturer's service manuals and courses, and in *Television and Video Engineers Pocket Book*; we shall confine ourselves now to fault-finding suggestions.

A complete lack of operation should lead to a check of supply voltages – to the tuner and to the ICs used in its control. If these are present, check that the reference oscillator (here crystal X1) is running, then that the tuner oscillator is running, indicated by an output from the prescaler, which may be internal or external to the tuner: in most designs the prescaler output frequency is within the range of an ordinary workshop oscilloscope. If these are in order examine the 'tuning' output of

the control chip, which consists of a square wave whose duty cycle varies to define the tuning point. It should be possible to see the pulse width varying as different stations are selected, or as the self-seek sweep is carried out. After integration this PWM waveform becomes the d.c. tuning control voltage, and the remarks made above apply.

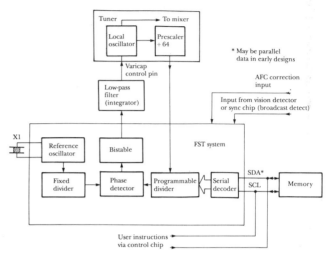

Figure 7.4 *Simplified diagram of frequency-synthesis tuning system*

If necessary, check that the clock and data pulses are being generated in the appropriate chip, and passed to the peripheral ICs (more of this in Chapter 9), and that tuning commands are reaching the programmable divider chip. Check, too that bandswitch programming is correct where relevant. Memory chip problems cannot affect tuning up/down functions, but can prevent storage and recall of a programme. First check clock and data interchange between memory and control chips.

Tuning drift is less likely with PLL tuning systems since the whole operation is geared to the reference clock crystal, typically running at a precisely defined frequency around 3 or 4MHz – tuning drift will result if this crystal wanders at all. When used with local sources of r.f. signals such as computers, TV games and VCRs, drift is more likely to arise in the 'transmitter' than the receiver – check that a.f.c. is available (i.e. programmed) on the channels assigned to such signal sources: the user's instruction book for the receiver gives details.

Self-seek tuning

Various forms of auto-sweep tuning are available, some incorporated in very early TV models. A problem sometimes encountered with these is failure to stop sweeping when a station is found. It is generally due to lack of a feedback signal to indicate the presence of a signal. Identify and investigate the feedback line, which may come direct from the vision detector, or more often from the line oscillator section which puts up a 'flag' when it becomes phase locked. If the line oscillator's free-run frequency is wrong, then, or if there is some other factor present to cause slow lock-in, the tuning sweep will pass each transmission frequency without stopping. For investigation, a fixed voltage can be supplied to the tuner to maintain signal receptions.

VCR r.f. boosters

To overcome the losses in plug/socket and r.f. splitting systems, VCRs incorporate an r.f. booster, generally a two-stage transistor amplifier between aerial-in and aerial-out terminals. These often fail or go low-gain – check, though, that any attenuator switch (marked local/distant/DX) is correctly set, and that the supply voltage is present and correct. If so, the usual cause of trouble is a faulty transistor, difficult to deal with, especially where the manufacturer treats the device as a 'black box', giving no circuit or spares data. Replacement modules are expensive, particularly where the r.f. modulator is also incorporated.

The faulty transistor can usually be traced by its odd electrode voltage readings – once the leadouts

have been identified, perhaps with the help of a transistor data book. Usually the base voltage is high due to an O/C b-e junction, and a replacement transistor will cure the problem at minimal cost: the replacement must be of a suitable type.

Response shaping

The i.f. response curve is defined by a SAW filter or by a bunch of discrete tuned LC circuits with a characteristic curve like that of Figure 7.5. If the response curve-shape is wrong, various problems arise, depending to some degree on how strongly other channels are picked up by the aerial. Some symptoms are: buzz on sound; low, noisy or no colour; poor definition of picture verticals; streaking of picture features due to poor l.f. response; etc. They are easily identified as being i.f. alignment problems by the fact that they change as tuning is adjusted. Where a SAW filter is used, it can only be changed; where discrete tuned circuits are used (now rare), alignment can be carried out by reference to the manufacturer's instructions, using a 'spot-frequency' method or – better – a wobbulator to give the response curve of Figure 7.5. Before doing this, however, check that vision demodulator and a.f.c. tuned circuits are correctly aligned on 39.5 (38.9)MHz: see below.

System conversion

One of the traps in the i.f. shaping filter is tuned to co-sound frequency – the 33.5MHz notch in the curve of Figure 7.5. When converting a receiver from system B/G to system I or vice versa this notch,

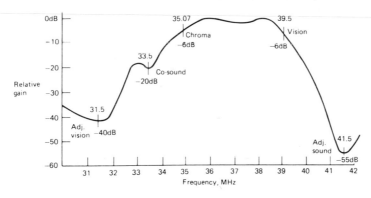

Figure 7.5 *Response curve of the i.f. filter, based on a vision i.f. carrier frequency of 39.5MHz*

and the adjacent sound one at 41.5MHz, must be moved to correspond – more details in the next chapter.

IF amplifier

The i.f. signal from the tuner must now be brought up to a level high enough for the demodulator to work, and a large reserve of gain is required so that the a.g.c. system can cope with a wide range of signal strengths. Occasionally a discrete i.f. amplifier like that of Figure 7.6 may be encountered, typically using a three-stage amplifier downstream of a response-shaping network with five or so pre-set coil adjustments. Faults in this type of circuit are fairly easily traced by making voltmeter checks of transistor electrode voltages, bearing well in mind the influence of the a.g.c. control circuit, which can act to upset operation. The simplest way to deal with this is to override the a.g.c. action by introducing a variable potential from a low-impedance external source. Here it would be applied across C114d, and − 1V would be appropriate. If the amplifier now starts to work, the a.g.c. control circuit is faulty: check the voltage at the signal sampling point (sometimes a faulty *video* amplifier transistor is responsible) then for the presence of gating pulses where relevant, then for faulty (i.e. leaky) transistor or diodes. Incorrect operation of the a.g.c. circuit can drive the i.f. amplifier into non-linearity, leading to such effects as poor sync (pulse crushing, Figure 5.2(a)) or soot-and-white-wash pictures. Failure of a.g.c. decoupling capacitors (and this applies to later IC circuits) can in particular upset field sync, the corrupt pulses being visible on an oscillogram of the detected video signal; see Figure 5.2(b).

If the amplifier does not operate when given correct a.g.c. potential, the problem will usually be due to faulty transistors or leaky ceramic capacitors, typically those decoupling the transistors' emitters. DC voltage checks at individual electrodes soon locate the problem. In Figure 7.6 the transistor's b-e junctions typically go O/C, betrayed by a high base voltage: for transistors T16d and T17d any base voltage in excess of 0.8V indicates a faulty device.

The early design of Figure 7.6 is unusual in using diode envelope detectors for vision and intercarrier sound, respectively D19d and D18d. Virtually all vision demodulators are now of the synchronous type as described later.

IC i.f. amplifier

Much more common are IC designs of i.f. amplifiers; they are sometimes incorporated in a 'module' with the tuner. A typical chip i.f. stage circuit diagram is shown in Figure 7.7. This sort of circuit is quite trouble free. Problems in and around IC51 are unlikely to cause a very noisy picture (except indirectly through the tuner a.g.c. line); a milky, flat picture, an over-contrasted effect, flattening of whites or poor sync are more likely symptoms of i.f. chip trouble, the key test point initially being IC pin 12, which should produce a well-defined video output of about 3V peak-to-peak. It is best examined at point 4/8, where the fuzziness due to the 6MHz sound carrier has been filtered out: L32 is adjusted for minimum fuzz on sync pulses. Incorrect or no video output should lead on to a check of the i.f. chip pin voltages, starting with the supply (12V at pin 11) and the a.g.c. voltage at pin 14. Zero voltage here, i.e. due to a S/C C41 or grounding of module point 4/9, will mute the chip and delete sound and vision output.

Strange effects, similar to those of i.f. misalignment, stem from incorrect tuning of the carrier coil L34 – it is associated with the vision demodulator. An alignment procedure for this is given in the service manual, and must be followed where teletext reception is incorporated. For less critical applications the coil may be adjusted for subjectively-best reproduction of a broadcast test card: optimum reproduction of detail and lowest picture brightness generally coincide at the correct point. Carry out this adjustment with a.f.c. switched off and the set carefully tuned. Figure 7.8 shows two possible effects of maladjustment of the carrier coil. With the a.f.c. now restored, adjust L36 for spot-on tuning.

The a.g.c. circuit operates as a loop around the chip, and emerges to control the tuner gain at module pin 4/1. For minimum picture noise the tuner is kept at full gain until the r.f. input level rises to the point where cross-modulation and non-linearity in the tuner is threatened, symptoms of which are sound-on-vision, patterning, picture shading or floating bars on the picture; also sound buzz. In these circumstances 'back-off' the a.g.c. crossover control VR36 until the effects disappear. Again, the electrolytic capacitors associated with the a.g.c. system are well worth checking for strange picture or sync defects; here they are C35, C36, and C47.

The pre-amplifier IC50 is provided to compensate

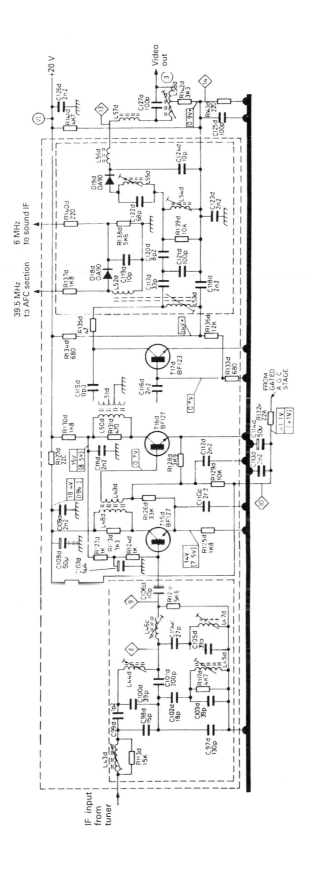

Figure 7.6 *IF amplifier using discrete components in filter and amplifier sections (ITT)*

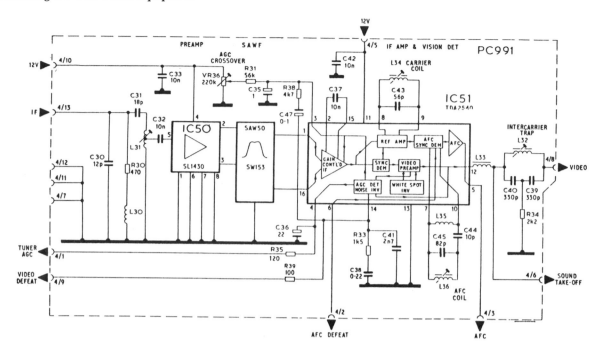

Figure 7.7   *I.F. processing system based on a SAW filter and single chip (Ferguson)*

for the losses introduced by the SAW filter SAW50, and to provide a differential drive to the latter. They give little trouble; any that does arise usually manifests as noise effects for a faulty chip, 'ghosting' or 'ringing' for a defective SAW filter – check its grounding and connections before suspecting it, however. The high frequencies (40MHz) and low signal levels (mV) in this part of the circuit means that the carrier envelope shape

cannot be displayed on any but a high-gain wideband scope – not the type normally found on the repair bench!

Full-function i.f. chips

Some late models of TV and VCR use a comprehensive IC which takes an input direct from the

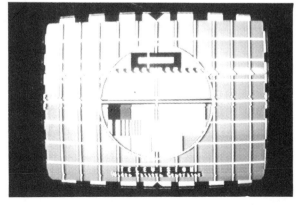

Figure 7.8   *These off-screen photos show the effects of mistuning of the vision demodulator coil. In (b) the sync is starting to be affected*

74

SAW filter and delivers demodulated vision *and sound* at its output pins, thus covering the subjects of both this and the next chapter. The remarks made above and in Chapter 8 about IC i.f. systems still apply fully. Extra features to be found in some of these designs include separate provision for sound i.f. processing, with a dedicated output from the SAW filter; noise reduction and mute circuits; and even incorporation, within the one chip, of sync separator, line and field oscillator, video processing and sound demodulation and amplification. Be guided by the manufacturer's block diagram and pin voltage specifications.

Intercarrier immunity

In tracing no-picture faults, do not be misled by the presence of a sound signal, which suggests that the vision i.f. stages are working correctly: the sound signal normally passes through the i.f. amplifier at very low level, and is almost immune to noise. Thus a sound signal can sometimes pass unscathed through a virtually dead amplifier.

Interference

Various forms of interference can spoil reception of TV programmes and tape replay. Where a clash occurs around Ch36, the modulator frequency of a VCR can be trimmed by a semi-accessible preset. Citizens-Band and other radio interference can often be suppressed by a low-pass co-axial filter fitted in the aerial socket. During times of high atmospheric pressure, *co-channel* interference can have the effect of superimposing drifting lines and bars on the picture, usually characterized by a prominent vertical and/or horizontal bar: little can be done about this except wait for weather conditions to change!

Other forms of interference can be more difficult to track down and diagnose. Check that the TV or video is not itself responsible by trying another, and bear in mind that a TV set placed too near a VCR can cause various forms of interference by direct radiation into the sensitive low-level replay amplifier circuits – check by physically separating the two. Some relief can often be gained by interposing a foil or metal screen between them. Interference from more distant sources, where they upset reception of *broadcast* TV and radio transmissions, can be investigated by the Department of Trade and Industry: the first step is to obtain the booklet *How to Improve Television and Radio Reception* from a main post office. In general, the later the design of the receiving equipment the greater its immunity to interference of all sorts.

8

AUDIO RECEIVER SECTION

This chapter will deal with the entire audio section of TV sets, and the 'audio tuner' part of VCRs to the point where a baseband audio signal is delivered to the sound recording section. As in the two previous chapters, we are dealing mainly with low-level circuits whose reliability is good.

System diagram

Figure 8.1 shows a two-chip receiver set-up, which is at once a block diagram and a circuit diagram. Only the left-hand section is used in VCRs; the rest of their audio sections will be dealt with in Chapter 17.

The demodulated video signal contains a strong component at 6MHz (UK system I) which corresponds to the transmitter's sound/vision carrier spacing, and which contains f.m. sound. This is filtered out by the 6MHz ceramic filter X100 at the left of the diagram, and passed to a multi-stage limiter to strip off amplitude modulation due to the vision signal. The clipped f.m. carrier is demodulated

within IC103 according to the 'quadrature' carrier feed developed across coil L110, whose adjustment is critical for buzz- and distortion-free sound reproduction.

Demodulated audio signal now passes into an electronic attenuator, whose output depends on the potential applied to IC pin 5, a point worth remembering even if the facility is not used, as here. The output from the attenuator emerges from the chip on pin 8, and re-enters, a.c.-coupled, on pin 4 for further amplification. The final product from pin 3 passes to the output stage – in this design via an ordinary volume control in the form of a potentiometer manipulated by the customer. Very often these types of volume control go 'noisy', crackling whenever moved. While this is often due to a worn control track, any d.c. current in the track exacerbates the situation, so C140 or its equivalent is worth checking for leakage before cleaning or replacing the track.

The audio output stage in Figure 8.1 consists of a simple IC with two heat-spreader tabs; the only unusual feature here is the method of coupling the

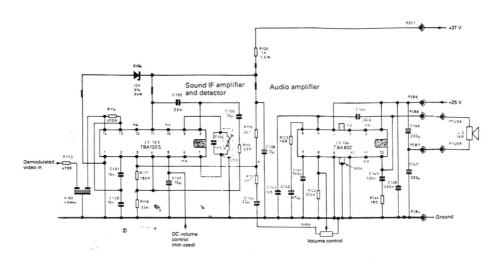

Figure 8.1   *Two-chip TV audio system (Decca)*

loudspeaker via two capacitors which also perform a supply decoupling function.

Fault finding

A complete absence of sound in an otherwise working receiver should first be approached by listening very closely to the loudspeaker. Complete absence of anything audible suggests that the speaker is O/C or disconnected (check any earphone socket or speaker switch); or that the supply voltage to the output stage is missing, perhaps due to failure of a fuse or safety resistor. Using an analogue meter on the lowest ohms range, a crackle should be audible from the speaker when its continuity is checked, and when (as a check of coupling capacitors etc.) the prods are connected between ground and pin 12 of IC104. If, alternatively, a close listening test reveals slight hiss or buzz from the loudspeaker, the output stage is working, and the fault further upstream. With the volume control set high, a finger on its slider connection should give a loud buzz from the speaker: if not, IC104 and its peripherals should be checked; if so, check IC103 signals at pins 8, 4 and 3 and the d.c. potential at pin 5. Again it must be stressed that these suggestions are applicable to all types of circuit once IC pin functions etc. have been identified.

Distorted sound is a very common problem. Often the loudspeaker itself is responsible, with an off-centre cone. If so, it shows up most at low volume settings, and can quickly be proved by hooking up a substitute speaker for test. If the distortion is still present in another loudspeaker, check that the audio section supply voltages are correct, then that the quadrature coil (L110 in Figure 8.1) is correctly adjusted. Incorrect adjustment of this coil can also be responsible for sibilant distortion with rasping 'Ss'; and *vision buzz* on sound. In a very few cases where the sound take-off filter (here X100) is in the form of an LC tuned circuit, its incorrect setting can have the same result of buzz on sound.

Premature or repeated failure of the output chip may be caused by excessive supply voltage (in the particular case of this circuit, a strong possibility is failure of the shunt regulator which controls the 25V line); by a too-low speaker impedance (faulty loudspeaker or something nasty in the speaker/phones socket?); or by inadequate heat sinking –

some fin-type heat sinks rely on physical pressure or on a layer of compound for heat transfer, and these may fail. The output chip may also be responsible for intermittent sound or cracking if the symptom is little affected by the setting of the volume control, but first check peripheral components, especially electrolytic capacitors. Components responsible for this sort of problem can often be identified with a squirt of aerosol freezer.

One-chip system

The greater simplicity of a one-IC audio system is demonstrated by the circuit of Figure 8.2, in which the entire audio section of a large-screen TV is handled in a single chip. The comments made above apply equally to this circuit, which also incorporates treble and bass controls, the latter operating in a negative feedback circuit. In this type of circuit the electrolytic capacitor decoupling the d.c. volume control line (here C533) can fail to give spurious oscillation effect, which may change pitch or disappear with volume setting, especially in remote-controlled sets. Again, beware the effect of any audio mute circuit – most modern sets have inter-station muting to suppress hiss while changing channels or tuning. The chip in Figure 8.2 is also notable for having audio-out and audio-in facilities, which in some designs are routed to AV connectors; see Figure 6.10. When external audio is being handled, the intercarrier channel must be muted.

Failure (O/C) of safety feed resistor R535 strongly suggests that the chip IC531 is faulty; before condemning it, check C541, C542 and C545, and see if current consumption returns to normal at zero volume control setting – if so, the loudspeaker is suspect. On rare occasions safety resistors like R535 here fail for internal reasons, a replacement permanently restoring normal operation.

In some designs the demodulator tuned circuit (here L531) is replaced by a special ceramic filter, different in type to that used for selectivity (i.e. CF531) thus removing the need for any trimming adjustments at all in the audio section. Some designs use two ceramic filters in series in place of CF531 for better selectivity. Very occasionally these filters (of either type) fail, resulting in buzz or distortion on sound. If the effect is intermittent, gentle heat and freeze treatment (see Chapter 20) may identify the faulty component.

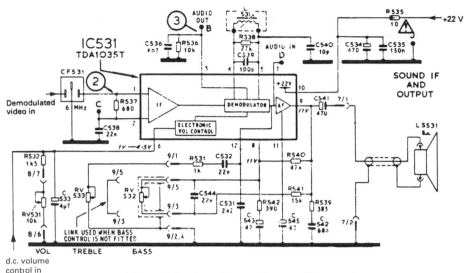

Figure 8.2 *One-chip sound system (Ferguson)*

Discrete audio output stages

Most TV sets now use IC audio output stages, though discrete types are still encountered: they generally use push-pull output circuits, configured in various ways, class A, class B complementary, and so on. When trouble develops, especially overheating or high current consumption, an approach which often saves time is to cold-check each semiconductor with an ohmmeter, replacing faulty ones as necessary. Figure 8.3 shows a class B circuit, using five d.c.-coupled transistors, where this technique is relevant. Excessive current through the output pair is indicated by the overheating of emitter resistors R409 and R410. This should direct attention, once any damage has been repaired, to quiescent current controllers R407 and T405, whose opening or high resistance causes high current in T403 and T404. If, alternatively, the path between the bases of the latter pair becomes too short (leaky or S/C T405; O/C track-end or maladjustment of R407) the result is crossover distortion, particularly noticeable at low volume settings. For this and the previously described audio stages the best way of checking for audio distortion is to inject a modulated audio carrier from a pattern generator and trace its progress with an oscilloscope.

In the circuit of Figure 8.3 the centre-point voltage of the output stage is set by a d.c. negative feedback loop via R403. Where the centre-point voltage is wrong the effect is clipping of the top or bottom of the audio waveform, heard as distortion at high volume settings. So long as the semiconductors are in order, the most likely culprits for this are R403, R404 and C75 and their equivalents in other circuits.

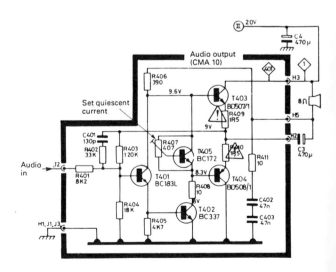

Figure 8.3 *Early design of audio output stage using a class-B transistor configuration (ITT)*

Standards conversion

TV sets and VCRs are sometimes presented for conversion to UK operation, having been purchased abroad. The project is best feasible if the equipment is already able to display good *pictures* as it stands. This means that it is capable of working on PAL signals, has a suitable tuner fitted, and that the modifications required are confined to those sections concerned with the sound carrier frequency. For UK/Ireland/Hong Kong System I this is 6MHz, while for other PAL areas of the world it is System B/G, 5.5MHz.

Primarily, the conversion consists of retuning the intercarrier sound i.f. channel up to 6MHz. Where discrete L/C tuned circuits are used (L110 in Figure 8.1, L531 in Figure 8.2, etc.), unscrewing the cores generally suffices, though a reduction in value of the parallel capacitor may sometimes be necessary. This tuning can be done by ear for maximum sound level consistent with minimum buzz and distortion, and the same applies to the 'selective' filter (equivalent position to X100 in figure 8.1, CF531 in Figure 8.2) where it is in adjustable (LC) form. The adjustment is best carried out on a transmission whose picture contains lots of peak white and sharp edges to give 'worst case' conditions for vision buzz: a test card or caption transmission is suitable.

Where (as is usually the case) the sound carrier tuned circuits consist wholly of ceramic filters, they must be replaced by 6MHz types purchased as standard spares for UK models – be sure to fit the right types in the right places. Where a UK equivalent model exists this is ensured by ordering the specific components required from the local spares agent. If not, the rule is to use SFE6.0 types for selectivity; CDA6.0 types for quadrature demodulation; and T6.0 types as notch filters in the vision baseband circuit. Figure 8.4 clarifies this, and also shows the positions of these filters. The vision notch filter (position C) may well be in discrete form – if so, tune it for minimum 'grass' on the line sync pulse as described on p. 57.

Once the intercarrier sound section has been retuned to 6MHz the audio will be reproduced correctly, but patterning may upset picture reproduction, especially in areas where adjacent TV channels are strongly received; and some buzz may mar the sound, especially with sharp, contrasty pictures. These problems arise from the fact that the i.f. filter's traps are not tuned to adjacent- and co-sound frequencies: to remedy this the SAW filter must be changed for a System I type or, in a circuit similar to that of Figure 7.6, the traps retuned 500kHz downwards. See Figure 7.5 and manufacturer's alignment instructions.

In the case of VCR conversions, it is also necessary to arrange for the *output* sound signal to conform to System I, easily done by retuning the sound carrier generator coil in the r.f. modulator up from 5.5 to 6MHz. This coil is sometimes difficult of access, but easily recognizable as the only separately screened coil in the modulator 'box'. Again, it should be critically adjusted on a sharp contrasty picture off-air (off-tape pictures are not sharp enough to test with) using a TV set known to be in good condition.

Conversion can also be made in the opposite direction: System I to System G. This is less often required, however.

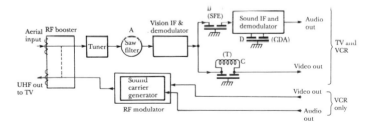

Figure 8.4 *Positions of filters etc. involved in systems conversion. The lettering is (A) co- and adjacent-sound traps; (B) sound selectivity; (C) notch filter in vision channel; (D) quadrature demodulator resonator*

9

TV CONTROL SYSTEMS

This is a chapter that would hardly have existed in a book on television a few years ago. The pace of development has been very rapid, to the point where now there are few TV sets (even portables) without some form of remote control. In sophisticated TV receivers there are probably more transistors and diodes working for the control system (though they are buried in chips) than in the rest of the set put together!

Apart from now-obsolete systems using ultrasonic sound waves as the carrier, TV control systems fall broadly into two categories: simple ones with few buttons on the IR handset and a 'decoder' chip at the receiving end; and advanced types with microprocessor control, memory capacity and provision for a teletext option in the set. In this chapter we shall examine one example of each of these, purely from a fault-finding and diagnostic point of view. The comments on IR remote control handsets, preamplifiers and command-decoders are equally applicable to VCRs.

Infra-red handsets

The sender unit is a very simple device so far as the serviceman is concerned – Figure 9.1 shows the circuit of a typical example. It contains primarily an encoder/pulse generator IC, and a minimum of peripheral components: a keyboard matrix, a pair of power LED drivers, and two infra-red emitter LEDs. The most common cause of trouble is a low battery: check it first! If the battery runs down quickly (a *minimum* of 6 months life can be expected with normal use) check the quiescent current, which should not exceed a few microamps. If it's excessive, leakage in one or more of the keyboard contacts is a likely cause, easily checked in many designs by disconnecting the keyboard altogether. If the high current consumption continues, remove the pulse drive by disconnecting R1303 or Q1301 base; if the current now drops, the IC itself is responsible so long as no signs of any

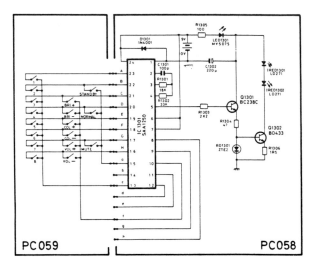

Figure 9.1 *A typical remote-control handset circuit diagram (GEC)*

liquid spillage are present on the PC board. Continued high current consumption necessitates disconnecting all possible culprits in turn until normality returns: take off IC1301 pin 24, D1301, and Q1301 and Q1302 collectors.

The next most common symptom, perhaps, is a complete loss of remote control, raising the question of whether the sender or receiver is responsible. It's essential to establish this before continuing! The simplest way to check for IR emission is to use a commercially available 'Magic Mirror' (see Figure 2.5) which reflects IR radiation in visible form, in reasonably subdued ambient light. An alternative IR sensor can be made in one of the forms given in Figure 9.2, and will also serve (if built into the plastic case of a discarded ball-point pen or similar) to check for IR radiation from the 'cassette lamp' of later designs of VCR. With any of these sensors experience soon establishes what constitutes a norm, and even permits individual checking of each of the two or three emitting LEDs in the handset. Many TV cameras, too, respond to IR

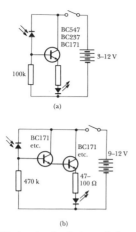

Figure 9.2   DIY circuits for infra-red detectors. The pick-up and emitter LEDs can be salvaged from scrap equipment. RS Components types 308-506 (sense) and 586-475 (emit) are suitable

light and will show the 'glow' from the front of the handset in a darkened room.

Where the emission from the handset is absent or weak, the first check (after the battery) should be for an O/C or dry-jointed reservoir capacitor – here C1302. The high pulse current needed draws on the reservoir rather than the battery in the short-term. Remote control handsets are more vulnerable than any other equipment to physical damage – by heavy-handedness, liquid spillage, being dropped, corroded by exhausted batteries, etc. Where the unit has failed, then, a close and careful inspection of the interior will often show the cause, be it a cracked PC track, O/C battery connector, or dry-jointed LED. Many designs have ceramic resonators, operating at about 450kHz, in place of the timing components C1301/R1302 of Figure 9.1; the resonator sometimes fails and is worth checking by substitution in cases of no oscillation, no pulse drive.

Possible causes of *weak* radiation are failure of one of the emitting LEDs; drive faults: check that Q1302 fully saturates at each 'on' pulse; and an accumulation of nicotine, grease or tar deposits on the infra-red filter window at either end of the IR link. Bear in mind, too, that strong ambient light (e.g. sunshine) can desensitize an infra-red control system.

The keypad works on a matrix system (details later) and usually depends on a rubber sheet with moulded-on 'key bumps' and corresponding 'contact pills' which bridge printed lands on a contact PCB. Erratic operation of the keypad, or misfiring of one

or more individual keys can usually be cured by a thorough clean of the PCB contact surface and the conductive pills of the rubber sheet. If in doubt, or if cleaning is not fully effective, replace the sheet. Often the start of the problem is liquid (coffee, beer, spirits) ingress, which can also have the effect of making the individual keys physically sticky: dismantling and cleaning is the cure.

In the rare case of incorrect commands, again check for leakage or damage to the PCB contact system before suspecting the IC itself. The problem here is that the fault may possibly be in the decoding chip at the receiver, and it is difficult to analyse the serial data issuing from the remocon chip: ideally, either a known-good substitute handset or TV set should be used to establish whether the encoder is faulty. The key-scan system used in IR handsets will be discussed in Chapter 13.

Command receiver/decoder

A circuit diagram of the simpler form of TV control system is given in Figure 9.3. The IR signal is picked up by photodiode D1851 (top right corner) and amplified in IC1851, a multi-stage preamplifier chip. A peak-to-peak output signal of several volts should be available at its pin 6. If not, decoupler C1855 could be responsible, but this particular IC, type TEA1009, is a more likely culprit; indeed it has been found faulty on many occasions.

The heart of this control system is the SAA1251 decoder chip, here designated IC1501. The serial data from the IR sender enters on pin 16, and can be checked at output pin 17 if required. Within the chip, the remote commands are decoded. Taking the analogue functions first, each of pins 2, 4 and 5 ouputs a pulse train whose mark space ratio is approximately 1:1. At a remote command for brightness-up, colour-up and volume-up respectively, the duty-cycle at each of these pins increases; for down-functions the duty-cycle decreases, and this can be clearly seen on a scope hooked to the IC output pins. The pulses are integrated by RC networks to form a d.c. control voltage whose level is proportional to m/s ratio. This is 'added' to the control voltage derived from the set's front-panel control. If the capacitor in the integrator (i.e. C1042, colour level) opens, bizarre effects result: alternate stripes of colour and monochrome in diagonal bars. Light and dark bars result from a pulse train on the brightness control line, and a raucous tone from the speaker in the case of the

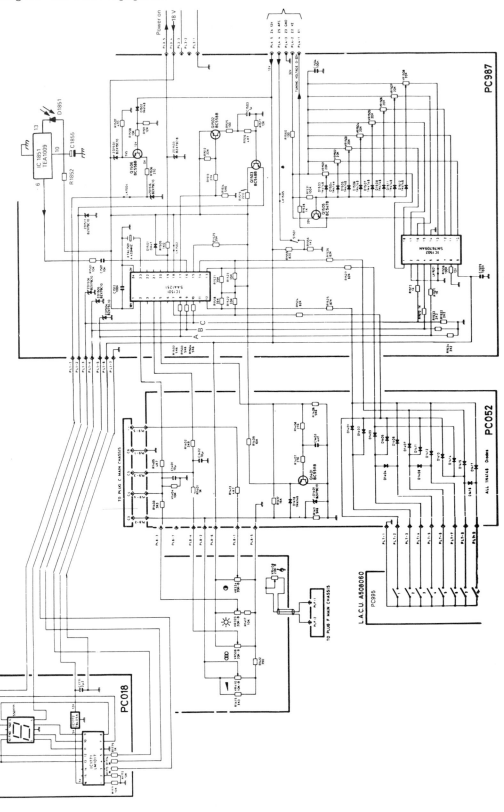

Figure 9.3 *TV control system using SAA1251 chip (GEC)*

sound control line. All these effects change with use of the remote analogue up/down features. At switch-on, or when the remote 'normalize' button is pressed, the mark-space ratio reverts to 1:1, and the levels of brightness, colour and volume return to those pre-set by the main controls.

The numbers entered into the remote keypad for channel selection are decoded to 3-bit parallel binary data on IC501 pins 8, 9, 10; 3 bits can describe eight states, and thus there is a choice of eight channels. The 3-bit parallel bus has two destinations: display driver IC1771, internally programmed to switch on the appropriate sections of the seven-segment channel display according to the data; and IC1502, the tuning switch, which decodes binary-to-decimal in order to ground the bottom of the required tuning pot, one of VR1501-8; see Chapter 7 and Figure 7.3.

These three lines, marked ABC in Figure 9.3, are the key to diagnosing problems in this and similar TV control systems, some of which have a fourth line D to increase the 'event description' to 16. If one of the lines is inactive, i.e. stuck high or low, wrong commands will be the result on all functions except those whose codes happen to agree with the fault condition. It is easy, with an oscilloscope or logic probe, to check for activity on each of these data lines, which are normally held low (in this particular case) by pull-down resistors R1532/3/4. A stuck line may be due to an internal S/C in any of the three chips connected, but first check passive (and inexpensive!) components like parallel capacitors, or here the 10V protection zeners ZD1504/5/6.

Where the parallel-data fault is obscure, get the set into a known condition (i.e. TV Channel 1) and check the status of each of the lines in the bus according to the setmakers' truth table published in the manual. This will identify decoding errors in any of the three chips IC1501, IC1502, IC1771; and using the front-panel-mounted channel-selection keys will help to tie down a wrong-command situation to the handset encoder chip or the main decoder IC1501. This is because channel selection from the front-panel controls is also made via the 'ABC' bus and IC1501, according to the states of the latter's pins 7, 12, 13 and 14.

Each of these pins is normally held high by pull-up resistors R1518/21/22/23 from the 18V line, and various combinations of them are grounded momentarily by the front-panel channel buttons via a diode matrix D1401 to D1417. A leaky diode here gives incorrect channel selection from the front panel buttons, but does not prevent correct selection via the remote handset.

The other remote control function handled by the circuit of Figure 9.3 is standby switching. The action is centred on pin 19 of IC1501, which is where the first check should be made in the event of remote standby switching problems. The standby command, entering as serial data, pulls pin 19 low, latching the PSU regulator chip off via Q1601 and Q1602 in the standby power supply section; see Figure 9.4. Failure to switch off is best approached by careful checking of each stage in this chain, ensuring that a full pulse is communicated to the PSU chip. At remote switch-on IC501 pin 19 goes high, kick-starting the main PSU via the action of

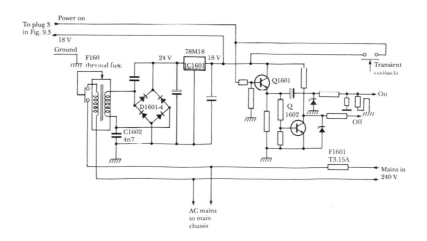

Figure 9.4 *Standby power supply for use with the circuit of Figure 9.3*

Q1601; again trace the 'on' pulse through each stage individually in the event of failure to come on, checking particularly any electrolytic coupling capacitors in the chain. At front-panel switch on, the same latch-on mechanism is invoked by an auxiliary contact pair in the mains switch, shown in Figure 9.4 – check it if switch-on can only be achieved via the remote handset.

Depending on detail design, loss of the standby power supply (here derived, see Figure 9.4, from a small mains transformer, rectifier bridge and 3-leg stabilizer) may prevent the set from working at all (as here) or just shut down the remote control section and possibly the front panel channel selectors. In this event, the stabilizer IC, here IC1601, type 78M18 is suspect; some designs use 12V and 5V regulators of similar type and these too are suspect. If voltage supplies are present, correct and reaching the main IC, check that its clock oscillator is running (XTL1501 in Figure 9.3, scope to IC501 pin 23) and that command data is entering at pin 16 before suspecting the IC itself. Some command-decoder chips use a ceramic resonator (about 450kHz) for clock generation, and this can intermittently or permanently upset operation by drifting in frequency or going O/C.

This form of control system may be found in some of the early Teletext receivers, typically with a command-translator chip in the text-interface panel, and with clock signals coming from IC1501 pins 21 and 18/22. The same fault-finding techniques apply, but it should be borne in mind that teletext commands are not *processed* by IC1501 – merely passed on from its pin 17. So long as valid serial data emerges here, then, fault-finding is confined to the interface section – generally on a sub-panel associated with the text decoder.

Full-facility remote control

Although serial data is necessarily used in the infra-red link of the simple system just described, the data are processed in parallel form within the set. Later and more sophisticated control systems use serial data within the set, basd on a two-wire system: one carries clock pulses (SCL) and the other coded data (SDA). These systems are characterized by an 'intelligent' control microprocessor, a RAM memory store for tuning and other data, and an interface chip (CITAC) for frequency-synthesis tuning and control of analogue functions. All these systems are text-ready, and the

handset has the capability, if required, of addressing and controlling many separate pieces of equipment. Again, we shall examine a typical system from a practical point of view, with trouble-shooting firmly in mind.

Microprocessor system

The essence of modern TV control systems is a central microprocessor controlling the rest of the set by means of bi-directional data on a two-wire serial data bus. Commands from the remote handset and from front-panel keys are decoded by the microprocessor and distributed around the set on the data bus, whose two lines SCL and SDA carry pulse trains ('data bursts') to instruct various ICs. The SCL (serial clock) line is held high when no commands are present; during each message it carries a squarewave for timing and synchronization purposes. The SDA (serial data) line also sits high during quiescent periods; during a message it carries a series of 8-bit data words. First in each word comes a start bit, signified by a drop to zero during a clock 'high' period. The first word is an address, identifying the receiving device, and is followed by an acknowledge bit, sent 'high' by the microprocessor but held low by the slave to indicate its presence and its reception of its own address. Next comes an address word to indicate the particular register within the slave device into which the data is to be written – typically an assigned *location* in the separate memory, the volume control register in the CITAC chip (see later) or the page memory section of the text controller. The third word on the SDA bus is the command data itself which conveys the actual message. This is decoded and actioned by the slave device whose output carries out the request, be it to change text page, increase colour, switch to AV input, change to the TV programme whose tuning data is held in memory location 3, etc. The slave chip thus acts as an interface between the control system and the set's signal-handling stages.

A generalized block diagram of this type of system is shown in Figure 9.5. The heart of the system is the MAB8441 microcontroller, whose inputs come from the infra-red receiver; the local keyboard; and the data on the 2-line I²C bus. Its outputs go primarily via the I²C bus, and to various devices connected directly to its I/O ports, pins 2–9 and 18–27. These are mask-programmed according to the setmaker's wishes at the time of IC

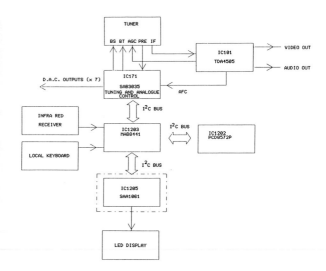

Figure 9.5 *Outline of control system using the Mullard/ Philips I²C bus communication system. In this range (Panasonic) only 'luxury' models use IC1205, SAA1061*

manufacture. The programs differ widely between different makes, so reference to the setmaker's data is essential when fault-finding in this area.

CITAC chip

The CITAC (Computer Interface for Tuning and Analogue Control) chip in Figure 9.5 is type SAB3035. Its four main inputs are serial data commands from the microcomputer; serial data from the memory IC PCD8572P; a feed from the prescaler within the tuner, typically fosc/256; and an a.f.c. error line from the i.f. section. Its decoded outputs are up to eight analogue controls for brightness, contrast, volume, etc.; a tuning voltage for the varicap tuner; bandswitching lines for the tuner; and such functions as sound mute, AV switching etc. as required by the setmaker. An internal block diagram for this device is shown in Figure 9.6. Faultfinding consists mainly of checking these inputs and outputs, initiated by commands from the remote hand unit.

Memory section

The stored data (channel numbers, brightness/ contrast/volume/colour levels, favourite text pages, special instructions per channel like AV time

constant and frequency offset, etc.) is held in the memory chip PC8572, which has 1K of non-volatile memory arranged as 128 × 8. This device (Figure 9.7), apart from supply and ground lines, has only two connections to the rest of the circuit: the SDA and SCL lines, with all address and data information passing in serial form along the former. A faulty memory chip does not prevent 'direct' functions like sweep-tune, standby switching etc., but will give rise to such symptoms as incorrect analogue levels at switch on, failure to store information, or false readout of 'stored' data, leading to incorrect pre-set tuning, etc. In cases of failure to memorize, before condemning the memory chip check that the user's 'store' command is reaching the main microprocessor and giving rise to data on the I²C bus. At 'normalize' or channel change command there should go onto the data bus from the microcomputer the addresses for memory chip and location, then the data should flow from the memory IC itself, to be acted upon by the CITAC.

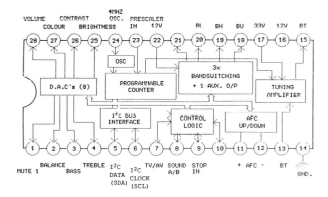

Figure 9.6 *Internal operation of CITAC chip SAB3035. The uses of the DAC output pins on the left depends on the individual setmaker, who arranges the system software to his own requirements*

Fault diagnosis

Because programming varies between setmakers it is difficult to give specific fault-finding information. It is not possible, with such ordinary equipment as an oscilloscope or logic probe, to *analyse* the data on SCL and SDA lines, though a burst of pulses should be visible briefly during message trans-mission. The general form of this is shown in

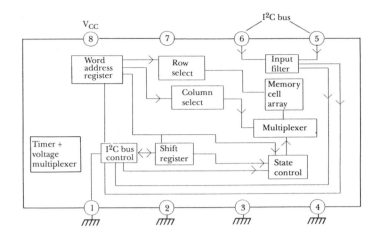

Figure 9.7   *Internal arrangement of the serial memory chip PCD8572. Several similar memory chips are available, all working to the same principles on the I²C bus*

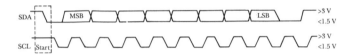

Figure 9.8   *Data format on the I²C bus*

Figure 9.8: use a double-beam scope and adjust the triggering carefully if it is required to observe the 'start' condition. A scope with a delay line in the Y-channel facilitates this. Data errors are rare, and with the odd exception of a faulty memory, if the data is present at all it's usually correct; if only one 'slave' chip is malfunctioning it is more likely to be faulty than the control microprocessor. Sometimes one of the I²C data lines can become inactive due to being 'stuck' high or low, generally as a result of one of the chips connected to it developing internal leakage or S/C: check by disconnecting each device in turn until normal conditions return, and bear in mind that IC failure can be caused by EHT flashover elsewhere in the set.

More often faults take the form of complete loss of operation due to an absence of response and data on the bus lines. Start at the microcomputer, ensuring that its clock oscillator is running, that it resets at switch-on, and that its power lines are present, correct to within 10% voltage-wise, and free of ripple or hash. Another fault symptom which may be encountered is random or incorrect functions at switch-on or on a sporadic basis. For this, check first for dry-joints and good pin contact of any pluggable ICs; for microcomputer reset at

switch-on; for hash/noise on IC supply rails and associated voltage supplies, typically 5V, 12V and 33V lines; and for a 'noisy' infra-red preamplifier output. The latter is most easily checked by disconnecting the pre-amp altogether: if the photo-diode or IR amplifier chip was responsible, normal operation (via front-panel controls) will return. This problem is also encountered on certain makes and models of VCR, some using a discrete IR preamplifier circuit.

Fault-finding charts

Figure 9.9 suggests a fault-finding procedure for inoperative controls, based on the most common systems in use. Figure 9.10 outlines a more general check on the control system; the 'conclusion' boxes should be taken to implicate the *functioning* of the IC or other device specified rather than its internal workings, so check peripheral components and operating conditions before suspecting the device itself. As a corollary to Figure 9.10 if programmes can be tuned by self-seek and memory, but not by direct entry of channel number, first check that the reference crystal (generally a 4MHz type

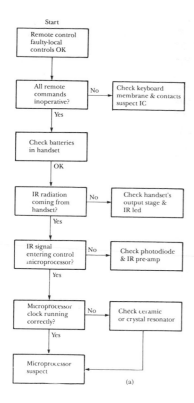

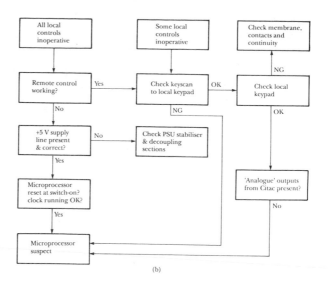

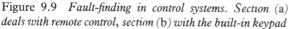

Figure 9.9 *Fault-finding in control systems. Section (a) deals with remote control, section (b) with the built-in keypad*

associated with the CITAC chip: pin 24 for TDA3035) is running at the correct frequency, then that the prescaler is operating correctly – by tuner substitution or with a frequency counter. If these are in order the CITAC chip is suspect.

Individual features

Some TV sets using microcomputer control make use of another control line called DLEN (data line enable) for such functions as display drive (typically 'luxury' models with separate drive chips to a comprehensive display panel) and certain teletext control systems. The DLEN line carries a framing pulse for the data and clock trains: refer to manufacturer's service manual.

Some TV sets and VCRs incorporate control features which can be misleading to the diagnostician especially when they are invoked for the

Figure 9.10 *Procedure for dealing with tuning problems in sets equipped with FS tuning and I²C communication*

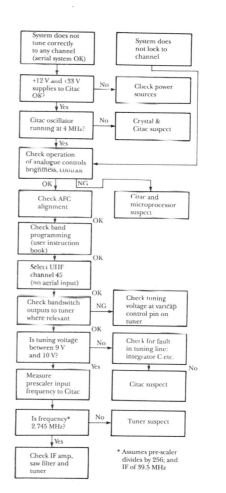

wrong reasons. Examples are *sound mute* and *inter-station blanking* facilities which may cut out sound and/or vision signals if the control section fails to get 'synchronization-locked' or 'signal present' messages from the timebase or i.f. sections respectively. Their absence may also invoke a 10-minute timer or similar artifice which switches the set into standby on the assumption that broadcast transmitters have shut down.

Control from peripheral devices

Finally, sets that incorporate AV sockets often have limited control facilities via the sockets themselves.

In the 6-pin DIN socket system, pin 1 selects signal routing as follows: for a TV socket all pins become outputs with pin 1 high, all pins become inputs with pin 1 low. The opposite is true of a VCR socket. In the 7-pin DIN system, pin 6 takes serial data for remote control, as does pin 2 in an 8-pin DIN interface. Scart systems (see Figure 6.10) have more comprehensive facilities: pin 8 provides simple on/off source switching for in/out signal routing, while pins 10 and 12 are assigned to serial control data, with access to equipment-internal buses.

10

PICTURE TUBES

The picture tube is the largest and most expensive component of a TV set or monitor, be it black-and-white or colour. As such, it is essential that fault diagnosis of a picture tube is correct: tube replacement is seldom economically justifiable outside the guarantee period unless a rebuilt tube is available and the rest of the receiver is in good condition.

Correct setting up is important too, though modern picture tubes and their drive circuits have few adjustments – very close tolerances in the manufacturing process has removed the need for most of them. The delta-gun colour tube has now disappeared from the domestic TV scene, and with it the banks of controls used for setting up convergence and geometry. Here we shall concentrate on its successors, the in-line gun types, whose development has been rapid.

Setting up

The main setting-up process for a picture tube concerns the ring magnets on the tube neck. Their design and arrangement varies somewhat between tube types, but the basic system is shown in Figure 10.1. There are three *pairs* of ring magnets, arranged thus: 2-pole, 6-pole, 4-pole, counting from the gun end of the tube. The rearmost pair (2-pole) is concerned with colour purity; the symptom of incorrect purity is a 'staining' effect on the

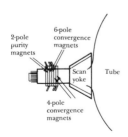

Figure 10.1   *Neck-ring assembly on a 90° PIL-type picture tube*

colour in some fixed area of the display. To adjust purity allow a 15 or 20 minute warm-up period, then manually degauss the tube. Slacken the securing screw of the scan yoke. Pull the yoke back 2cm or so, and arrange for a red raster to be displayed. Carefully adjust the 2-pole ring-magnet pair for the widest and most central area of red you can achieve (Figure 10.2). When the deflection yoke is pushed forward again, the red area should expand to cover the whole screen area; clamp the yoke in the middle of the range over which it has no effect on the pure red raster. Check the other colours: they should be pure over the whole screen area, and the white raster (all guns restored) should now be free of tinting or staining. *Problems* with purity will be discussed later in this chapter

Figure 10.2   *Screen display during adjustment of purity*

Convergence

The front two ring magnets in Figure 10.1 form a *4-pole* pair, and they are concerned with static convergence: the registration (overlay) of the centre-points of the three individual colour images which form the composite picture. Before we discuss their operation it's necessary to define the gun arrangement in use. The earliest in-line gun tubes had the red gun central, and the ring magnets manipulated the green and blue (outer) beams to conform to the red image. Most of these tubes were 'factory-sealed' types, in which (in theory, anyway) subsequent adjustment of the ring magnets is forbidden. Later types of tube have the *green* gun central, and the object of the convergence set-up is to make the pictures traced by the red and blue guns coincide with that from the green gun.

The 4-pole rings are adjusted, by differential- and co-rotation, to overlay the red and blue centrepoints of a crosshatch or dot display to form a magenta spot. This may not coincide with the green centrepoint: if not, the third ring pair (6-pole types, see Figure 10.1) should be adjusted to align and overlay the green and magenta images at screen centre.

The achievement of correct static convergence will in most picture tubes also render acceptable dynamic convergence, meaning that the red, green and bue images coincide and overlay at all points on the viewing screen. In some designs of in-line tube, provision is made for a trimming adjustment of the scan yoke to optimize convergence at screen edges. These tube types can be recognized by the fact that the front of the scan yoke has a larger flare than the tube it partners, so that with the rear of the yoke clamped some panning and tilting of the deflection coil assembly is possible. While carefully watching a crosshatch pattern in a mirror, manipulate the yoke to achieve good edge- and corner-convergence: pan horizontally to overlay red and blue lines parallel to and adjacent to screen edges; tilt vertically to overlay the extremities of red and blue lines which pass through the screen centre.

When optimum convergence is achieved, wedge the yoke in position with the rubber wedges provided by the tube-maker, then glue all in place, including the purity and static convergence rings. Figure 10.3 shows the wedges in position, as well as the other neck components mentioned so far. With

modern tubes, no convergence error anywhere should exceed 2mm, with 3mm just acceptable in the screen corners of 'difficult' tubes.

The first type of in-line gun tube produced by Mullard/Philips involved a thick-neck system with completely separate gun assemblies; many of them are still in operation. The arrangement of the neck magnets is slightly different here, incorporating internal gearwheels to ease adjustment. The effects of the rings are as already described, but the order is different, as shown in Figure 10.4. The procedure here is to move the yoke fully forward by means of the yoke axial position lever, then adjust the purity ring lever (4th magnet tab from gun end) for best red purity, as described above. Now withdraw the yoke again by the axial position lever for good overall purity.

Next the 4-pole rings (second and third tabs from gun end) are adjusted for centre convergence of red and blue images, and then the 6-pole pair (rings furthest from the gun) for overlay of green and magenta images, as described earlier. There is one additional ring and tab on this multipole magnet assembly: this, the rearmost one, has a two-pole field like the purity ring, and is a *geometry* control, adjusted to remove any 'bow' effect on the horizontal centre-line of the picture.

The deflection yoke in these thick-neck multi-gun tubes is only adjustable in an axial direction. To set screen-edge registration, half a dozen preset *dynamic convergence* controls are provided, most working through a four-pole correction coil built into the yoke assembly. If there is no response to adjustment of these, check that the four-pole coil is connected. Sometimes correction in one direction only is lost, generally because of loss of a pulse feed from one or other of the timebases: a typical cause may be a dried-up (O/C) electrolytic capacitor in the field sawtooth feed to the convergence circuit.

Trinitron tubes

While the basic principles of the Sony Trinitron tube design have always been the same, the methods of setting up vary widely with tube size and vintage. The purity setting is as already described, with the additional need on some wide angle (114°) types to affix small disc magnets to the tube bowl to correct small areas of impurity. The adjustment of horizontal static convergence (R/B verticals at screen centre) is made by a potentiometer associated with the EHT supply: for most sets it's

Figure 10.3 *Tube neck components. The yoke-positioning rubber wedges are clearly visible on the glass bowl*

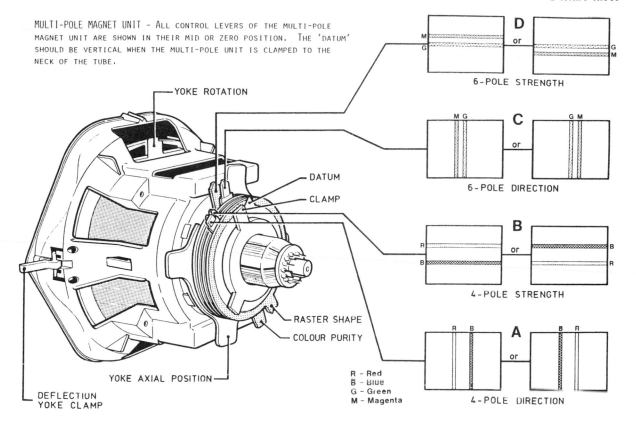

Figure 10.4 *Mullard 20AX deflection yoke, neck rings and the effects of their adjustment*

mounted on the EHT rectifier block; in later models it's on the tube base panel. Further correction of h-stat convergence, if required, is made by adjustment of the BMC magnet, nearest to the tube base (Figure 10.5).

Vertical static convergence (R/B horizontal lines across screen centre) is aligned by adjustment of the V-stat control magnets, also shown in Figure 10.5. If large adjustments of these static convergence magnets are required, check the purity again – it may need re-setting.

Like the other types of in-line tubes, this design of tube and yoke has no *inherent* dynamic convergence error, but some tubes require 'trimming' to take up manufacturing tolerances. In Trinitron tubes this may be done in one or more of the following ways:

1 Yoke tilt/panning and wedging, as already described.
2 Providing adjustable field- and line-rate waveforms in a 4-pole coil mounted just behind the

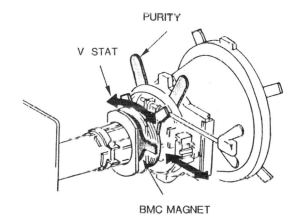

Figure 10.5 *Neck-magnets on a Sony Trinitron tube*

deflection yoke: amplitude and tilt controls are generally provided. Consult the service manual for the model in question.

3 Fitting small disc magnets to the tube's glass

envelope near the deflection coils, a trial-and-error process for correction of small convergence errors in the screen corners.

Most Trinitron tubes have a co-axial EHT connector, the outer conductor of which carries the EHT voltage, and the inner the convergence control voltage, about 450V lower. Incorrect connection of the inner conductor has strange effects on the picture, i.e. an all-green screen or gross misregistration of red, green and blue images horizontally, with the h-stat control having no effect.

Grey-scale tracking

The object of the grey-scale tracking process is to achieve correct and true colours in the display, without any tinting or bias. By adjusting the bias and drive levels to each cathode individually, the three electron beams are held in step, intensity-wise, all the way from black level to full drive, ensuring a colour-balanced picture at all brightness levels. First the d.c. levels at the R, G, B cathodes are adjusted by individual presets (labelled bias or black level) to equalize the guns' *cut-off* points. On a black-and-white display (ideally a step-wedge from a pattern generator) there will now be no tinting in the darkest areas of the picture. Next the highlight (brightest) areas of the picture are examined for any sign of colouration. If such exists, careful analysis will show which of R, G, or B is deficient or dominant: adjustment of the appropriate drive control – a preset contrast control fitted to each colour output stage – will remove the colouration to give pure white highlights, again on a black-and-white step-wedge pattern. At least one drive control (that for the 'weakest' gun) should be left at full gain. If difficulty is experienced in achieving correct grey-scale, it may well be that one or more of the guns is suffering from low emission – see below.

Tube fault symptoms

The most common problem encountered with picture tubes is low emission. The effects (Figure 10.6) are low brightness and defocusing, particularly noticeable in a colour tube, where the afflicted beam is one of three which are critically balanced and matched. A low emission monochrome tube often develops a 'silvery' effect in the white areas of

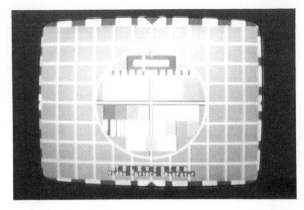

Figure 10.6 *This shot was taken from the screen of a low-emission tube, and shows defocussing and 'flattening' of whites*

the picture. Depending on the hours of use, and to a lesser degree on the tube type and manufacturer, ageing becomes noticeable after 4–7 years of operation. In sets fitted with auto grey-scale (e.g. using TDA3652A, see Chapter 6) the effect will be masked until the emission of the gun concerned falls below the circuit's correction range, when the picture will deteriorate very rapidly. In sets not so fitted, periodic adjustment of the grey-scale tracking is required to compensate for falling emission unless all three guns deteriorate at exactly the same rate.

When the emission becomes very poor it is no longer possible to achieve good grey-scale performance, and defocusing occurs at high brightness levels on one or more colours. To get a true picture of tube performance, link all three cathodes together (this should do no harm so long as the input signal is a monochrome one) then the picture displayed shows a direct comparison between the guns. Tubes can be checked and often reactivated by means of a rejuvenation machine: see p. 14.

Inter-electrode leakage

Occasionally, electrical leakage develops between the electrodes in one of the gun assemblies. Invariably this turns the affected gun hard on, to flood the screen with light of the colour generated. Modern tubes have all electrodes, except the cathodes and heaters, common to all three beams, so that leakage of the cathode turns on an individual gun, while leakage at other electrodes turns on all three, probably invoking some overcurrent protection device via the heavy EHT load. The resultant pumping or shutdown effect only takes place when the picture tube has warmed up, and disappears if the tube base is unplugged. Some tube testing machines (see p. 14) can clear inter-electrode leaks. A symptom similar to electrode leakage results from the tube's grid being allowed to 'float' electrically due to a cracked print or similar problem on the tube base. It is safe, however, to permit the *cathode* to 'float' in this way. Disconnecting a cathode from its feed circuit turns the gun off.

Purity problems

Adjustment of purity (beam landing) has already been discussed as part of the setting-up process. Beam landing is upset by magnetic fields in the vicinity of the tube, so it's necessary after tube replacement to degauss the tube and its magnetic environment with an external degauss coil. This mains-powered device is passed closely over the tube, then withdrawn to some 2m distance before switching off. Its action is also useful for demagnetizing any nearby ferrous objects which may upset purity – examples are a steel radiator or a rolled-steel joist, which latter may be concealed in a wall. Sometimes impurity is caused by a Hi-Fi loudspeaker placed too near the TV set, or even by replacing the set's own speaker with an unsuitable type. TV speakers have low stray magnetic fields.

If purity gradually deteriorates over a period of days or weeks, only to be restored by use of an external degaussing coil, it's almost certain that the set's internal degauss system has failed. A quick check is to feel the temperature of the degauss posistor at switch-off after a run; it should be very warm to the touch. *Cycling* purity errors, where the 'stain pattern' slowly moves or changes in intensity, must stem from a mains-rate magnetic field near the tube: check the auto-degauss control components for leakage; disconnecting the circuit from the mains supply forms a quick check. If the problem remains, check any mains transformer or choke for excessive current or faults.

If the picture becomes impure on bright scenes, the cause is expansion and distortion of the shadowmask due to high wattage dissipation. First check that the beam limiter circuit is correctly operational, then check the purity setting in case it is 'borderline'. Sustained high brightness in a small area of the picture (i.e. certain teletext patterns) can cause local overheating and impurity even in a correctly working tube and set.

An incorrect or off-tolerance setting of the purity rings can cause the picture to be off-centre in the screen, a point worth checking where no shift controls are provided. Normally a picture tube's tolerances are close enough to maintain picture centring within 2 or 3mm of true, depending on tube size.

Focus problems

Most 'poor focus' symptoms arising within the tube are due to low emission and ageing, as already described. If adjustment of the focus control takes the picture through an optimum focus point (albeit bad) check the EHT voltage and tube heater voltage before suspecting the tube itself. If the focus remains poor throughout the range of the focus control, first check the tube base components associated with the focus voltage feed: a series resistor may have failed, or electrical leakage be taking place at the spark-gap or at the tube base pin, usually visible as a blue discharge in a blacked-out room. Sometimes corrosion takes place at these points, visible on careful inspection, and cured by replacing the socket, spark gap or base panel as appropriate. When fitting a new spark gap be sure to use the right type for the tube: some are rated at 5kV, some at 8kV, and all are crucial safety components.

The same faults may be responsible for intermittent defocusing or random white blobs on the screen, and the mini-discharges can cause random channel changes, tripping or other odd symptoms in sets using sensitive ICs.

Where good focus cannot be achieved at any setting of the focus control, use a suitable (low current drain) meter to check the focus voltage. If it is not specified in the manual, early colour tubes require about 4.5kV, later ones (i.e. Mullard 30AX) about 6.8kV, and modern mini-neck high-performance tubes over 7kV. These figures are necessarily approximate! If the voltage is low, disconnect the focus lead from the base panel and measure again. If the new reading is higher, there is electrical leakage in the tube or on the base panel. Unplug the tube to check which. If, alternatively, the new reading is still low, check the focus voltage *source* once the EHT voltage level has been checked OK.

Physical faults

Some faults in picture tubes are due to physical causes. Most of them are fairly obvious, so we shall not dwell too long on them. Occasionally one or more of the mask holes may be blocked, creating a small black hole in the picture. This is often due to loose debris, especially in rebuilt tubes, and a slap on the faceplate with something like a slipper may well dislodge it. More serious is the rare situation of a displaced shadowmask: purity cannot be set, and a 'twang' or rattle may be heard when the tube is moved or tapped.

Sometimes a tube may go 'soft' which means that its vacuum is impaired. The effect is poor focusing and/or a 'flat', milky picture, perhaps accompanied by a visible soft blue glow inside the neck. If the vacuum is lost altogether (i.e. seal broken, etc.) the heater shows continuity and current, but does not glow, and there is usually violent sparking between tube electrodes if the set is switched on. Very occasionally neck fracture takes place below the yoke clamping ring due to too tight a clamp pressure or sparking between the clamp and a tube-internal electrode. In the latter case the cause could be excessive EHT voltage: check it when the tube is replaced.

It can happen that the faceplate becomes damaged in transit or as a result of user-abuse. If the faceplate is chipped or cracked in any way the tube must be changed. Only surface scratches (the sort capable of being 'polished out') pose no threat to safety.

All the defects mentioned in this section render the tubes unacceptable to rebuilders.

Flashover

Internal flashover sometimes takes place in the first few weeks' operation of a new or rebuilt tube. The sharp crack is disconcerting to viewers, and the energy liberated can destroy semiconductors in various parts of the set. To contain the discharge a carefully arranged circuit (Figure 10.7) uses spark gaps at every electrode, with a short link between the spark gap ground ring and the tube's outer conductive coating. The electrical circuits are grounded to the tube base panel only, not to the graphite coating on the bowl, or the rimband. If flashover does damage the set, check these arrangements, and try to ascertain whether the flashover

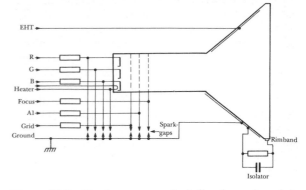

Figure 10.7 *Spark-gaps are wired directly to the tube's outer conductive coating. Series resistors absorb flashover energy to prevent damage to the set's electrical circuits*

really is in the picture tube or in some component like an EHT tripler, focus unit or LOPT overwind. If so, the earth path arranged for the tube spark gaps will afford no protection for circuit components. Excessive flashover, inside or outside the tube, should lead to a check of the EHT voltage, which may be high due to too high an HT line voltage, or incorrect tuning of the line output stage – see Chapter 4.

Tube biasing

Many faults that upset tube operation are due to incorrect voltages on electrodes. If the cathode 'floats' or goes high (i.e. drive stage cut-off) the gun turns off. Similarly, a loss of accelerating voltage (zero or low A1 potential) shuts down the gun. Too high an A1 potential, typically due to an O/C or high resistance in the 'earthy leg' of the supply network, causes an over-bright picture, usually with field flyback lines as shown in Figure 10.8. Reducing the brightness by some other means – like a pre-set brightness control – disguises the fault, but may leave the flyback lines visible on the picture.

If the cathode voltage drops, picture brightness increases, and where the RGB amplifiers' operating voltage disappears completely the tube is turned very hard on. Too high a grid voltage has the same effect, but this is rarer. In this sort of situation, prevent the risk of damage to the tube, EHT supply etc. by stemming the flow of beam current: disconnect the tube heater or unplug the base connector. In all cases of *no* picture, first check that

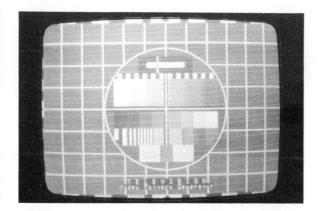

Figure 10.8 *The effect of too high voltage at the tube's first anode*

the tube heater is lighting up. This can save a lot of time! If it is, quick electrode voltage checks will establish the cause of no-picture or incorrect-brightness faults.

Handling picture tubes

Before replacing a tube, or connecting a tester to it, *discharge the EHT*, and leave a shorting link in place on the tube throughout the job. The atmospheric pressure on the outer surface of a large tube amounts to several tons, and the effect of an implosion can be devastating. For safety, follow these rules: wear gloves and goggles or a face mask; hold and carry the tube two-handed with the neck pointing upwards; do not chip, scratch or strike the glass; put the tube face down on a soft clean surface; leave the old tube in the TV cabinet until the replacement is to be fitted, and box up the old one as soon as it is removed; do not pick up a tube by its neck; and if possible avoid changing picture tubes in the field.

Tube salvage

An economic solution to the problem of a worn-out tube is to replace it with a rebuilt one. They are fitted with new guns and evacuated and sealed at high temperature. Often the safety-crucial rimband is replaced as well. Make sure that the replacement is an exact equivalent of the original, and that it comes from a reputable company. Run the set on soak test for a day or two to be sure that no flashover problems occur.

Usually the rebuilder requires the original tube on a part-exchange basis. Remember that it will only be acceptable if it's physically undamaged.

11

TELETEXT DECODERS

Apart from the first signal-processing stage, teletext decoders are wholly digital systems working at low power levels, and as such do not have a high failure rate. This and the fact that the decoder (as a handy pluggable module) has often been treated as a 'black box' and exchanged rather than repaired means that little information on servicing them has been generated. Sometimes a service-exchange or reconditioned module *is* the best solution to a breakdown problem. If diagnosis and repair is attempted, however, it is not necessary to have an in-depth knowledge of every detail of the workings of the teletext system in order to be successful, especially as all the circuitry is contained in a handful of LSI chips.

What *is* important is to have a clear idea of the role and function of the decoder itself, of the chips within the decoder and the key connection points between them. To this end a block diagram of a widely used form of text decoder is given in Figure 11.1.

Decoder overview

The video signal enters the VIP (Video Input Processor) chip at pin 16. The main blocks inside

this analogue IC, SAA5030, are as follows:

1 A sync separator whose field output leaves on pin 13, and whose line output is used internally to generate an FLR (Fast Line Reset) at pin 3.
2 A data slicer, whose 'squared' video-signal output emerges at pin 19 to enter the TAC (see later), and which is used to shock-excite the tuned circuit connected to IC pin 21. The clock run-in signal 10101010 at the beginning of each data line rings this LC circuit at 6.9375MHz rate to form a data-clock pulse train for use in the decoder section of the TAC chip.
3 A phase-lock loop, working on a sandcastle input at pin 5, which locks the crystal at pins 8/9 to 384 times incoming TV line rate. The ÷ 384 stage is in the TIC chip (see later) whence comes the sandcastle pulse to pin 5 to complete the loop. The locked 6MHz squarewave output at pin 6 forms the *dot clock*, the basic timing reference for the characters traced out on the teletext screen.

TIC

The next IC in Figure 11.1 is an SAA5020 TIC (TimIng Chain) device. Its function is to generate

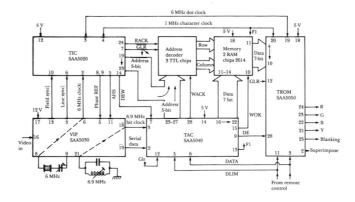

Figure 11.1 *Functional block diagram of the first type of mass-produced teletext decoder*

various timing and gating pulses for use elsewhere in the decoder. These are based on the 6MHz dot-clock signal from VIP, entering on pin 6. This is processed as follows:

1 A buffered output goes from pin 3 to the TROM (character generator) IC to time the basic text 'pixels'; the same pulse is divided by six to emerge as 1MHz at pin 4, also routed to the TROM as a character clock. Each character is six pixels wide. This 1MHz clock is also passed into the TAC chip (see later) for use in decoding remote-control commands.

2 Further division of the 1MHz clock by 64 renders a line-rate (15.625kHz) pulse, locked to off-air sync, which emerges as GLR (General Line Reset) at TIC pin 7. It enters TROM at pin 12 to start each line readout, and TAC at pin 12 for use in timing control.

3 A ÷ 625 stage renders 25Hz (frame) and 50Hz (field) rate pulse trains for use in an internal sync generator, whose AHS (After Hours Sync) emerges at pin 5 for use when no off-air syncs are present to time the text data already acquired and held in memory. The phase-locking sandcastle pulse leaving on pins 8, 9 is also generated from the same divider chain.

4 The various pulse trains locked to the 6MHz clock are applied to an internal decoder which generates several outputs: a DEW (Data Entry Window) output to gate on the text-decoding process during the field blanking interval (output pin 14, to 7 of TAC); a five-bit row address from pins 19–23 for the data RAM; and a RACK (Read Address Clock) at pin 24 to clock data into the memory system. The memory is written into during DEW, and read out during the active field period.

TAC

The third IC in the text decoder chip-set is the TAC (Text Acquisition and Control) device, type SAA5040. This is the heart of the text section: it carries out the decoding of text data signals. It has two control inputs conveying the user's commands – DLIM on pin 5, and DATA on pin 6. The main inputs, however, are serial data on pin 2; and the process-timing pulses: 6.9375MHz data clock on pin 3 and DEW on pin 7. The processes carried out in TAC are as follows:

1 The text serial data entering at pin 2 is enabled whenever the DEW line at pin 7 goes high. It is clocked into a serial-to parallel shift register under the control of the data clock pulse, pin 3. The data passing through this register is continually checked for a certain pattern corresponding to the *framing code*. When the correct framing code is detected the rest of the IC comes into operation, slicing the text data at correct intervals to form coherent 8-bit words.

2 Errors are corrected by use of a Hamming check for the vital row address data, and a parity check for the display data. So long as only one bit of the byte is incorrect the errors can be corrected; more than one bit in error leads to rejection of the byte. The data is continuously available at the seven bit port (pins 23 to 27) which leads to the memory – it is not written into the memory, however, until the WOK (Write O.K.) line (pin 15) goes low.

3 The WOK line is enabled when valid data becomes available, signalled by a match between the transmitted page number and the page requested by the user via control pins 5 and 6; the data for that page is written into memory, at locations determined by the data on the address bus (pins 23–27) and the WACK (Write Address Clock) pulse train emerging from pin 28.

4 Remote control pulses entering on the DLIM and DATA lines, pins 5 and 6, are decoded within the TAC chip and the data they contain (page number, time code) stored in a small on-board RAM.

5 A DE (Display Enable) line passes from TAC pin 5 to the character generator TROM to 'light up' the text display as required.

RAM store

The arrangement of the data RAM memory varies between setmakers. The system shown in Figure 11.1 is the most common in this type of decoder, and consists of two 1K × 4-bit static RAMs, type 2614. These are arranged in 32 × 32 matrices, an architecture which requires a ten-bit binary code on address pins 1–7 and 15–17. To interface this with the address-generators in the TAC and TIC chips one adder and two counter chips are used. The data is stored in the RAM memories in the same form as it is transmitted – 7-bit ASCII code (Figure 11.2) with one memory holding four bits of each byte and the other the remaining three. 960

Figure 11.2 *ASCII code for text character and graphics generation*

bytes (40 columns by 24 rows) are required to store a page of text.

TROM

The final IC in this text decoder chip-set is the TROM (Teletext Read-Only Memory). It is pre-programmed with all the alpha-numerical characters and graphic block-shapes required for text display as shown in Figure 11.2, and reproduces them according to the attribute code given in columns 0 and 1 of the figure. Internal functions and key pin numbers are as follows:

1 Data from the memory enters the TROM on pins 4 to 10 during the display period as defined by the RACK (Read Address Clock) from the TIC chip. Inside the TROM the ASCII decoder determines the character required.
2 The characters are generated as parallel data within the chip and converted to serial data for readout onto the TV screen under the control of the 1MHz and 6MHz character- and dot-clocks entering on pins 20 and 19 respectively. Outputs to the RGB interface are RGB signals on pins 24/23/22; a Y signal (monochrome text) on pin 21; a blanking signal (mutes the *picture* decoder outputs) on pin 25; and a superimpose output at pin 2, whose purpose is to reduce picture contrast during mix mode. All these are 'on-off' signals rising from 0V to 5V approx.
3 Remote control signals from lines DLIM and DATA enter the TROM on pins 11 and 3 respectively. Via a decoder within the TROM these control such user display commands as text/picture switching, superimpose and reveal.

Decoder fault finding

The test equipment required for text decoder servicing need not be exotic: a multimeter for d.c. voltage checks, an oscilloscope or logic probe for pulse checks, and occasionally a frequency counter are required. Faults encountered fall into three main categories: no text; scrambled display ('garble-degook'); and control faults. We shall deal with each in turn, based on the diagram on Figure 11.1.

No teletext

Where no text is displayed the first things to check are the power supplies to the decoder: 12V at VIP pin 17; 5V at TIC pin 12, TAC pin 14, TROM pin 18 and the memory chips – here pins 18. Check also that operating voltages are reaching the RGB interface, whose design varies between manufacturers. Of course an interface or text-board fault can prevent display of ordinary *analogue* pictures if the PAL decoder's ouputs are blocked on their way to the picture tube. Failure of power lines is generally due to a missing feed from a chopper or line output transformer, or a faulty series regulator in transistor or IC form.

If power supplies are present confirm that a CVBS video signal is entering VIP at pin 16, then that bit-clock (pin 21 then 18) and dot-clock (pins 8/9 then 6) are present. If not check the appropriate tuned circuits, then suspect the VIP chip.

In the presence of the clock signals there should be some sort of text-related display on the screen. If not, check the DE line (high at TROM pin 28 to enable text display).

Failure of the user's commands to reach or activate the text decoder can cause no text. So long as the other remote commands (for TV operation) are working, check that serial data is reaching TROM pins 11 and 3, seen on a scope as 'blips' at the moment of remote key-stroke. If not, it is possible that the 'text' button on the handset is faulty (check with Magic Mirror, see p. 18) but more likely that the problem lies in the control interface system: the remote control decoder chip or any I-bus translator chip. Service for this sort of circuit is described in Chapter 9.

Display errors

The most common problems associated with teletext take the form of display errors of one sort or another sometimes called 'garbledegook', the effect is shown in Figure 11.3. It can be caused by many things, often outside the text decoder module altogether. If the errors take the form of missing or incorrect characters at random, changing and/or improving with each page update or new page selected, the cause is usually data corruption. It may arise from a faulty aerial installation, where short-term signal reflections (nearby objects, poor cable dressing, faulty distribution system, impedance mismatch) can overlay adjacent text data pulses to cause decoding errors. The quickest and most practical test for this is to check with a substitute text receiver.

Problems in the receiver section upstream of the

Figure 11.3 *'Garbledegook' effect of corrupt text data*

text decoder can cause similar effects, and the resulting small degree of 'ringing' or phase error will not necessarily be visible on ordinary pictures or test card. While the tuner or SAW filter could be responsible, very often slight maladjustment of the vision demodulator carrier coil or a.f.c. tuned circuit is the root cause; see Chapter 7 and Figure 7.7.

To check that an adequate signal is reaching the decoder, the simplest method is to substitute the complete decoder module if one is to hand. If not (as is usually the case!) an eyeheight check can be carried out as follows: connect an oscilloscope to the decoder's vision input. Check first that the line sync pulses are square-cornered and that colour-burst amplitude is equal to that of the line sync pulses. Now trigger the scope's sweep at field rate, ideally from field timebase flyback pulses. 'Zoom' in on TV line 20 (a delayed-sweep scope with high EHT voltage is best), the second half of which contains a series of pulses at text-data rate. So long as the scope triggering is good and accurate the corresponding line on the odd field (no. 333) will overlay line 20. Line 333 contains complementary pulses to those on no. 20, opposite in polarity but otherwise identical in shape. The resulting display, then, contains a series of 'eyes' with fuzzy areas at top and bottom and a clear area between them containing only vertical lines. This clear area is called the eyeheight, and the worst eyeheight indicates the margin for error in the decoding process. An eyeheight corresponding to 75% of sync pulse amplitude is good, and should give completely error-free displays; an eyeheight equal to 35% of sync pulse amplitude is borderline, and likely to lead to data corruption.

Where *random* decoding errors take place in the presence of an input signal with good eyeheight the problem is in the decoder section itself: check that the 5V supply line lies between 4.8 and 5.2V, and that it is free of spurious pulses, noise and hash; if not, voltage regulators or decoupling (L, C) components may be faulty. Check also the grounding connections to the decoder module, then the capacitors at pins 23 and 24 of the VIP before suspecting the VIP IC itself. Sometimes random text errors are due to maladjustment of the bit clock (6.9375MHz) tuned circuit at VIP pin 21, a situation shown up by the fact that the errors tend to increase towards the right-hand side of the text display. Follow manufacturer's instructions in aligning the coil, and bear in mind that no effect will be seen on the display until the page is updated or reselected. The best page to show this type of error is the 'clock-cracker' engineering test page.

A second category of display error arises within the decoder, not this time due to errors in the decoding process, but to incorrect storage, readout or routing within the decoder. These problems are characterized by the fact that the errors, while appearing to be random, do not change or disappear when the page is reselected or updated: check on the main index or clock-cracker page. If incorrect characters always appear at a specific row/column position one of the RAM memory cells is faulty; similarly, if only one character is wrong (e.g. 'L' always appears as 'M') a memory cell is stuck – replace the appropriate RAM chip. If a great number of characters are wrong (again, always the same errors) the likelihood is that one of the data lines (TAC 16–22, TROM 4–10 in Figure 11.1) is inactive. Now all characters whose code agrees with the fault state (1 or 0, 'stuck' high or low) of the data line are reproduced correctly, while all other characters are wrong. By analysing the codes for the incorrect characters with reference to Figure 11.2 it is possible to determine which of the seven bits is incorrect, and thus identify the faulty line: in practice it is quicker to examine each line in turn with a logic probe or oscilloscope to find the 'stuck' one. Disconnection of each chip in turn, starting with the memory ICs, will then locate the S/C or whatever. Bear in mind that stuck data lines can also affect attributes (columns 0 and 1 of Figure 11.2) to give strange colour displays, monochrome text, graphics instead of characters, etc.

Stuck address lines generally have the effect of repeating groups of rows or columns; the repetition of the header row all the way down the screen, however, is generally due to failure of the RACK pulse train: check for a stuck TLC line due to failure of TIC or TROM. Again, there is little point in attempting to analyse the errors to tie the fault to a specific address line. A stuck one is very quickly discovered with a logic probe or scope, and diagnosed by isolating IC pins etc. until the line is freed. In the rare cases of two lines stuck together (address or data) a dual-beam scope will show them to be carrying exactly the same data at all times, and again IC pin isolation (and where necessary use of an ohmmeter) will provide the diagnosis. For this particular test, rock-steady triggering of the scope is important: use external triggering from FS and GLR lines for field and line sync respectively.

On-screen fault indication

Since the TV screen display is locked to line and field sync, and to the dot and character clocks, it forms a ready-made display system for all the digital signals in the text decoder. Where the interface takes the form shown in Figure 11.4, pulses may be displayed (with the receiver in MIX mode) by connecting the cathode of a IN4148 diode to the

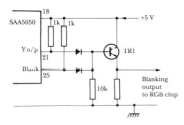

Figure 11.4 *Commercial interface circuit for text fast blanking*

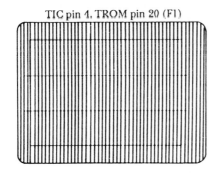

Figure 11.5 *Screen test display of 1MHz character block*

base of text blanking transistor TR1 and using the diode's anode as a probe, connected in turn to various key points in the decoder. Figure 11.5 shows the effect of the 1MHz clock; Figure 11.6 the memory address lines; Figure 11.7 the RACK/WACK; and Figure 11.8 the data lines to the SAA5050 TROM, whose effects (in the form of vertical black patterns at the top of the screen) vary when data like HALT/TIME/PAGE SELECT are keyed. Some very interesting patterns in the form of chequerboards and zig-zags are produced around the address decoder chips, types 74LS83A and 74LS161, but since these vary with decoder design only examples are shown in Figure 11.9. If regular work on one particular type of decoder is envisaged, it's worth noting and recording the characteristic patterns arising from the pulse trains at individual chip pins.

The pattern-tracing method described above is applicable to other designs of interface, following the principle of injecting pulses into the blanking/Y output section in MIX mode. To prevent damage to the TROM (open-drain outputs with external pull-up) ensure that either an isolating diode is present as in Figure 11.4, or that a resistor of say 1kΩ is present between the probe-diode and the TROM output pin.

Three key waveforms in the decoder (scope display) are given in Figure 11.10.

Miscellaneous text faults

An unsynchronized text display suggests that the 6MHz dot clock is not phase-locked to the incoming line sync – possibilities here are the VIP and TIC chips, but first check the 6MHz crystal, the components (especially capacitors) connected to VIP pins 1 and 7, and the continuity of the PLL loop between VIP pins 5 and 6 and the TIC chip. Other VIP and TIC faults generally give rise to no text operation, i.e. a blank screen. Failures of the TAC chip, where commands are entering it via the I bus (pins 5 and 6) generally cause inability to select text operation or pages, but before condemning the IC, check that GLR (pin 12) and F1 character clock (pin 13) are entering it.

TROM failures occasionally cause incorrect characters or attributes (always the same ones), but this chip more often develops a short at an output pin, sticking it high, open or low. For the RGB outputs this is fairly obvious, losing one colour of the text display or suffusing the screen with one

101

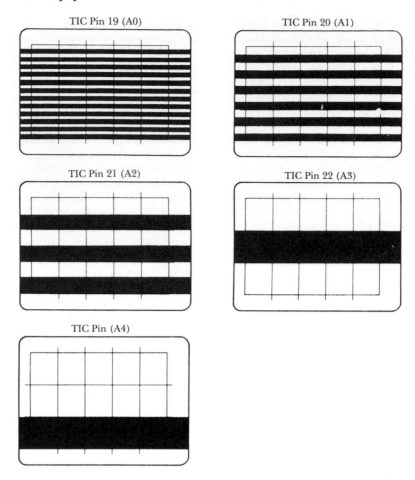

Figure 11.6 *Memory address line pulses displayed on TV screen*

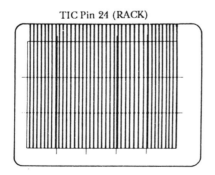

Figure 11.7 *Read/write address clock line display*

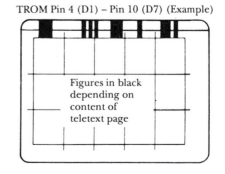

Figure 11.8 *TROM input data displayed at screen top*

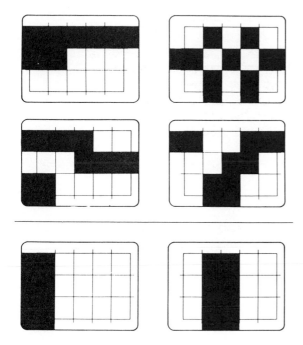

Figure 11.9 *Examples of address decoder chip pulses – their form depends on individual memory-bank configuration*

colour, R, G or B. Failure of the blanking output permits picture to intrude on the text display; if this pin sticks high, 'text only, no picture' will result. Repeated failure of TROM (or other text decoder) ICs is generally due to high-voltage flashover elsewhere in the set, or to excessive supply voltage, e.g. a S/C series regulator device on the 5V rail.

3-chip text decoder

A later generation of teletext sets uses a more modern design of decoder in a system called CCT (Computer Controlled Teletext). In concept this is very similar to the design already described though it uses three chips plus a memory, as shown in the block diagram of Figure 11.11. As before, the first chip is largely analogue in its operation, and operates from a + 12V line; the others are wholly digital LSI devices with no significant peripheral components. They operate from a +5V VCC/ VDD line. The interfacing is simpler, too, taking advantage of the later chroma decoder chips, which incorporate 'text' RGB and blanking inputs; and the I²C control bus, which here operates via a

separate Teletext Microprocessor IC for control commands.

A simplified outline of the functions within the CCT section, and the key interconnections between them, is given in Figure 11.11. To give an idea of possible culprits for VIP problems, many of the peripheral components are shown. The dot-clock PLL occupies the upper two links between VIP and CCT chips, and these should be checked wherever the text display loses synchronization; the filter networks on pins 19 and 21, and the 6MHz resonant circuit between pins 18 and 20 are also relevant to this. To set the free-running frequency of the dot clock, either remove the input signal (i.e. tune to a blank channel) and adjust the trimmer at pin 18 for 6010kHz at pin 17; or ground VIP pin 22 and adjust for steadiest display.

For 'garbledegook' check for a good quality 1V pk-pk video signal at VIP pin 27, then test the capacitors hanging on VIP pins 3–8 before suspecting the chip itself, but bear in mind the advice given above for the 4-chip decoder. It is just as relevant to this one.

The main outputs from the VIP IC are (continuing down the centre of Figure 11.11) the dot-clock at pin 17, sync at pin 25, serial text data at pin 15, and text clock at pin 14. Check these in turn.

The second IC (CCT) in the diagram combines the functions of TIC, TAC and TROM in the older design. More than half of its connection pins go directly to the RAM memory chip, which has a capacity of 2K2 for conventional text reception, 8K8 for *Fastext* in which several related pages (at the choice of user or broadcaster) can be stored for instant display. The eight data lines D0-D7 occupy CCT chip pins 22–29 and the address lines pins 30–40 then 2–3. A useful test pin on the CCT chip is no. 8, on which the quality of the text-data received by CCT is indicated: when a text line is received with no Hamming errors and correct framing code this pin goes high. It is reset for each line.

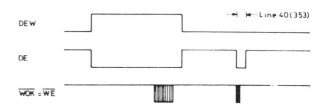

Figure 11.10 *Key decoder waveforms. For test-point identification see Figure 11.1*

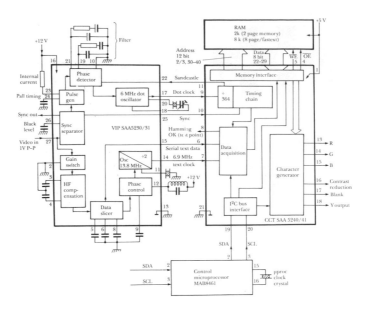

Figure 11.11   *CCT teletext decoder, with emphasis on key functions and interconnections*

The control chip (Text microprocessor) may not be found in the simpler text sets, in which CCT is directly controlled by SDA and SCL lines from the set's main control microprocessor. Here the dedicated text micro receives and issues commands for text operation.

Fault-finding

Virtually all the advice already given for the 4-chip decoder is relevant to this one. For no text display first check for 12V and 5V supply lines, video in, clocks running; then the data lines from VIP to CCT. For 'organized garbledegook' check for 'stuck' data and address lines between CCT and memory chips, isolating chip pins in turn on any afflicted ones to determine which IC is responsible. Occasionally sections of the RAM chip go faulty.

Problems with control operation can arise from a faulty CCT chip, but are more likely to lie in the functions of control lines SCL or SDA. In the configuration of Figure 11.11 the first check for faulty control operation should be the control microprocessor: look for operating voltage (+5V) and clock operation (here 4MHz crystal at pins 15/16); also loading or sticking of SCL and SDA lines before suspecting the text control microprocessor chip. In replacing this chip take great care that the replacement is obtained from the setmaker. The suffix to the type number indicates its 'firmware' program, customized for the manufacturer and TV model.

Odd faults encountered in the CCT chip by the author and his colleagues include odd colours missing (i.e. no green); no text, just a header line; and a strange garbled text display split down the middle by a blank vertical bar due to a line-phasing error.

12

VIDEO DECK MECHANICS

By their nature, electronic circuits (especially in VCRs) are generally reliable. The mechanical section, on the other hand, is subject to wear, damage and contamination. It has to work to very close tolerances and maintain them for years. For these reasons faults in and around the deck section account for a very large proportion of VCR service jobs, and so justify a long chapter in this book. Here we shall confine ourselves to the mechanical aspects of the deck assembly; the effects of failure and wear of the signal-handling parts of the deck (heads, servo pulse-sensors etc.) will be dealt with in their appropriate chapters, and the operation of deck-status sensors in the chapter devoted to system control.

Because failures of deck components can give rise to symptoms similar to those of syscon faults, and similarly tape-running problems, wear, etc. often result in picture and sound faults easily mistaken for electrical circuit problems, it is important to establish at the outset whether a symptom is due to mechanical or electronic failure. In general an oscilloscope is the best diagnostic tool for this. Throughout this and the following chapters guidance will be given; refer also to the symptom index at the back of the book.

We shall start here with a trip round a deck, following the tape past each component on its path, and examining them from a purely diagnostic point of view. For this purpose, Beta and Video 8 decks are similar in nature to the VHS type we shall be taking as our example, though differences will be examined at the end of the section. To aid identification of parts, Figure 12.1 shows a typical deck with all significant parts annotated. The numbers will be referred to throughout this chapter.

Threading

The first action of the machine in play or record mode is to thread the tape around the deck. Complete failure to do this, where fast-forward and

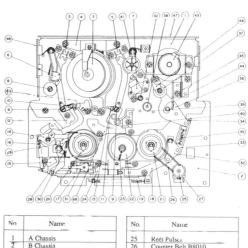

No	Name	No.	Name
1	A Chassis	25	Reel Pulser
2	B Chassis	26	Counter Belt B8010
3	Drum	27	Reel Sensor
4	V-Base	28	Shifter Arm
5	V-Blocks (T, S)	29	Auxiliary Brake
6	F.E. Head	30	Auxiliary Brake Spring
7	AC Head	31	Cassette Down Switch
8	S.I. Roller	32	Pinch Drive Lever
9	Guide Rollers (T, S)	33	Pinch Reciprocating Lever U
10	S Slant Pole	34	Pinch Reciprocating Lever L
11	T Slant Pole	35	Pinch Engagement Spring
12	Pole Base A	36	Pinch Roller Lever
13	Pole Base B	37	Pinch Roller
14	Tension Arm	38	Capstan
15	Tension Band Ass'y	39	Capstan Holder
16	Tension Arm Spring	40	Cassette Lamp
17	Shifter A	41	TI Roller
18	Shifter B	42	Stationary Guide
19	Shifter Spring	43	Capstan Motor
20	Supply Reel Disk	44	Open Angle
21	Take-up Reel Disk	45	Dew Sensor
22	Reel Idler	46	Entanglement Prevention Arm
23	Reel Idler Engagement Spring	47	Mechanism Intermediate Plate
24	Cassette Down SW Holder	48	DPG Intermediate Plate
		49	Back Tension Pole

Figure 12.1 *The mechanics of a VHS video deck. The ringed numbers are referred to throughout the chapter, and correspond to those on Figure 12.3 (Sharp)*

rewind functions are also inhibited, is generally a syscon action due to failure of the cassette lamp or the cassette-down switch. So long as these two are in order and operating power is available, threading should take place. Sometimes the action is accompanied by a loud banging and clicking, indicative of broken pinion teeth in the drive system – check by inspection. In older machines a slowing-up of the loading action, accompanied by sounds of distress from the loading motor (and possibly slippage of the loading belt where relevant) will usually be due to the loading pole bases 12, 13 binding in old, hard grease; or to a tight spot in the bearing of the loading mechanism. Remove any dried-up grease on the loading arm path with cotton buds and solvent, then replace it with a better lubricant – thin graphited grease, or as directed by the VCR manufacturer. If necessary these things can be checked by operating the loading mechanism by hand. No undue friction should be encountered.

If during threading the tape slackens and 'bounces', check the efficacy of the 'soft brakes' which bear on the spool turntables.

As the loading arms approach their end of travel the strain on their drive mechanism increases, and it is at this point that any loading belt is liable to slip if it is worn or stretched. The threading action may *appear* to be complete, but close inspection (and listening) reveals that the motor is still running and the loading-end switch not triggered. After a few seconds of this the tape is unthreaded and the machine reverts to stop mode. Sometimes temperature and other factors take a hand to render the effect intermittent, a typical symptom being occasional failure to make an unattended (timer-set) recording. Check for excessive friction in the loading mechanism and replace the loading belt.

Tape threading mechanisms work on the basis of a pair of contra-rotating pinions or sectors on the underside of the deck, and their mechanical phasing (and that of the pinions or sectors which drive them) is essential to the correct operation of the deck, especially in machines incorporating a single 'mechaposition' switch. If the loading arms are not in phase the result is 'crunching' of the tape as the cassette is lowered onto the deck; partial loading at inlet or exit side of the drum with loss of top or bottom of the replayed picture; or complete failure of one arm to engage the tape, resulting in a 'V-wrap' which only embraces about a quarter of the head drum periphery. In some later machines, and many camcorders, tape loading is done by the capstan motor 43, often along with other tasks like pinch-roller (37) engagement, brake manipulation, cassette-cradle release and even front-loading of the cassette cage. At every stage of checking and reassembly the mechanical components (pinions, sectors, slide-bars etc) must be phased and aligned according to the manufacturer's instructions and the alignment marks (holes, dots, arrows) in each individual component and in the mechaposition

switch, which is the effective governor of the deck's electrical drive system. A typical mechanism of recent vintage is shown in Figure 12.2 with the alignment marks correctly set. An incorrectly phased mechanism or switch generally causes:

1 Failure of the deck to operate in one or more modes:
2 Deck shutdown after an attempt at a mechanical function.
3 Oscillation of the mechanism, at switch-on or on user-request of a function, culminating in VCR auto-switch-off or reversion to stop respectively.
4 A graunching or clicking noise from an over-loaded and tooth-jumping pinion, worm or belt.

These things can be caused by a faulty mechaposition switch, but this is rare: it can be checked by

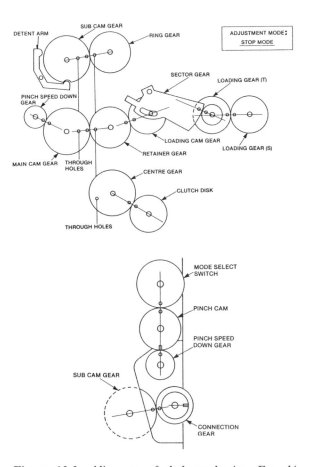

Figure 12.2 *Alignment of deck mechanics. For this Panasonic machine,* (a) *shows the below-deck arrangement and* (b) *the top-deck members*

analysing the data from the switch (double-beam oscilloscope and manufacturer's data) or by a substitution test on the switch itself.

Where a deck develops a fault which causes shutdown or mechanical strain, or where breakdown occurs with the tape guides in 'loaded' position, diagnosis can be made easier by driving the threading motor gently with an externally derived voltage. By this means the entire threading/unthreading (and eject where relevant, i.e. camcorders) process can be cycled or pulsed at will, checking mechanical behaviours at each step, and noting any excessive mechanical loading on the motor as reflected in its current consumption: workload and current are proportional. Beware of driving a prolonged and heavy current through a stalled or 'load-ended' motor, and disconnect its regular voltage drive (remove one wire) before applying the externally-derived voltage.

Tape path

A 'linear' representation of the tape path of a VHS machine is given in Figure 12.3. It largely represents Beta and V8 formats too, and many remarks made here are equally applicable to them. The prime mover of the tape is the capstan 38, which pulls the tape through the entire system shown here. The first event is the emergence of the tape from the cassette's feed spool whose turntable 20 must be at the correct level to avoid crinkling and damage to the tape. It's set with a gauging block and fit-as-required shims below the spool turntable.

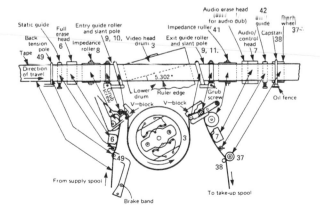

Figure 12.3 *VHS tape path, relating the 'plan view' to the linear sequence of deck components*

Back-tension regulator

The first deck-member encountered by the tape is the back-tension regulator pole 49. In most machines it acts as a purely mechanical feedback system to apply friction to the feed spool and maintain a specified degree of tension to the tape in its journey around the head drum 3. Two adjustments are generally provided: a position setting for the arm 14 and a tension setting involving adjustment of the anchor-point of the felt-lined friction band 15. They are carried out according to the manufacturer's instructions, by one of three methods: use of a *back-tension cassette gauge* with direct-reading scale in grams; a spring-loaded and hand-held tension gauge which hooks to the end of the tape on a dummy spool and is pulled by hand through the tape path; or a *Tentelometer* gauge through whose three legs the tape is threaded to give a direct reading on the instrument's dial. A typical back-tension specification is 35g-cm for VHS machines, 45g-cm for Betamax types. To convert between *torque* and *tension* units use the formula: torque (g/cm) equals tension (g) multiplied by reel radius in cm. Thus 30g tension is equal to 90g/cm torque for a reel 3cm in radius.

Some machines with direct-drive reel motors use an electronic control system for back-tension regulation. Here the measuring arm is linked to some form of electronic transducer and the 'pull-back' effort of the reel motor is set by a pre-set control in the reel servo circuit.

Regardless of how the back-tension is established or controlled, its correct adjustment is crucial to the correct operation of the VCR and to good recording and playback. Excessive back-tension causes premature head wear; misregistration of colours on pictures from tapes recorded or played back elsewhere; and in extreme cases can stop the tape moving altogether, perhaps on an intermittent basis. Insufficient back-tension can cause 'noise' at picture top or all over the screen due to poor penetration of the head tips; tape snatching or looping at the tension lever, particularly in cue mode; and colour misregistration as described above. A quick check for correct back-tension can be made by reducing the picture height on the monitor TV and examining the vertical picture components above and below the head-switch point of a known-good tape. A display like that of Figure 12.4(a) indicates insufficient back-tension; excessive tension renders the effect shown in Figure 12.4(b), while if the 'wobble-effect' is in both directions

(Figure 12.4(c)) the *dihedral* adjustment of the heads on the drum is incorrect. Some machines (mainly Beta types) make provision for dihedral adjustments: see the service manual.

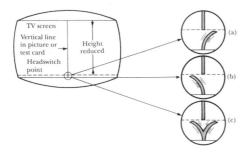

Figure 12.4 *Behaviour of picture verticals around the head switch-point: (a) low back-tension; (b) high back-tension; (c) dihedral fault in head-mounting*

FE head and impedance roller

The first 'fixed' component encountered by the tape is a static guide, adjustable height-wise. This one (not shown in Figure 12.1) is not critical, and is set by using a special gauging block – or, if necessary, by inspection for smooth, wrinkle-free tape running. After this the tape passes the full-erase head 6, whose gap is longer than the tape width, making vertical placement uncritical. The main requirement of the FE head from a physical point of view is that it be truly vertical, so as not to bias the tape up or down in its passage.

Not all machines are fitted with the impedance roller 8 shown in Figure 12.2. When it is present, it may come before or after the FE head, and its purpose is to smooth any fluctuations in the tape tension, which can give rise to horizontal jitter in the picture, especially at its top. In service the main things are to ensure that it is clean, undamaged and very free to rotate.

Tape guides

The tape is now approaching the head wrap, during which it will be guided and supported by the rabbet – an inclined 'shelf' machined into the lower drum assembly. At entry and exit points two components are required to position the tape correctly on the rabbet throughout its traverse of the head drum: the tape guide 9 and an inclined pole 10, the one to

set the tape level and the other to bias the tape downwards in order that it bears firmly on the rabbet. In VHS machines the guides have rotating sleeves of nylon, minimizing tape friction, but occasionally giving rise to an odd horizontal jitter or tear effect on the picture, which in bad cases is accompanied by an audible squeal from the afflicted guide. It is due to the sleeve becoming 'dry' on its shaft, and can be cured by dismantling and polishing the shaft, perhaps adding a few *specks* of graphite powder before reassembly. If in doubt, replace the entire guide assembly. Do not squirt oil into this bearing!

Correct guide alignment is vital. If the adjustment is a long way out tape crinkling and damage occurs. If it is a little out, picture reproduction suffers, with noise bars in one part or another of the picture. Eliminate all other possibilities for incorrect tracking or tape-running before attempting guide adjustment unless the guide(s) has been replaced or is obviously loose. Guide adjustment is carried out by screwing the entire assembly up and down, and takes place in three stages. The first is a rough setting for approximately correct height, made with the aid of a scrap tape or a gauging-block produced by the VCR manufacturer. The second is a careful adjustment made while closely observing the tape surface at the guide and at the lead-in of the head rabbet: first eliminate any crinkling or 'bubbling' of the tape, then screw down the guide until the tape is *seen* to just touch the end section of the rabbet on the lower drum assembly. The third and final stage requires the use of an oscilloscope, hooked to a test point in the pre-limiter stage of the replay f.m. amplifier. See the service manual for details. Adjustment is made for the most level and square-shouldered reproduction of the 'envelope' signal from a test-tape. Figure 12.5 shows the effects of various degrees of maladjustment in entry and exit guides. For a steady display synchronize the oscilloscope from the head flip-flop waveform (often called SW25) in the servo circuit.

VHS tape guides have locking screws on the guide-shaft itself or on the adjacent support block. During adjustment keep them sufficiently tight to prevent the guide rotating with the passing tape; when correct guide settings have been achieved, tighten them fully with the aid of a small hex wrench.

Inability to achieve good tracking without the tape crinkling or bubbling at guides or rabbet generally means that the tape is being biased up or down by some component *beyond* the afflicted

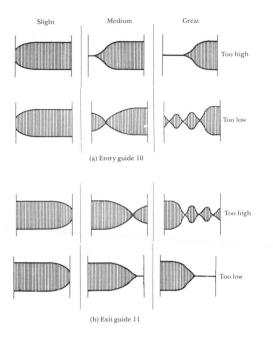

Slight　Medium　Great

Too high

Too low

(a) Entry guide 10

Too high

Too low

(b) Exit guide 11

Figure 12.5 *Envelope patterns displayed on an oscilloscope connected to the heads' f.m. output signal. Entry guide faults show on the left of each pattern, exit guide faults on the right. Too low a guide setting causes tape damage at the ribbon's lower edge as it buckles on the rabbet*

Head drum, 3

Most head-drum problems have to do with wear or damage to the head-chips themselves, manifest as picture faults. From the 'mechanical' point of view of this chapter, troubles occasionally encountered are as follows:

1 Excessive friction in the peripheral surface of the drum (possibly brought on by long running with too much back-tension) causing the tape to be picked up and 'billowed out' by a VHS machine during threading.
2 A *tilted* drum, caused by dirt, flux or foreign bodies between a newly-fitted upper drum and its baseplate, or by uneven tightening of its fixing screws. This causes an inability to achieve correct tracking by *both* heads at any one setting of the tracking control. On rare occasions, this is due to a faulty head-drum assembly, whose head chips are mounted at different heights.
3 Eccentric mounting of the drum, causing the head-tip penetration into the tape to vary throughout the tape-scan as described above. In severe cases this may be more noticeable as flutter of the replayed audio as the tape tension at the audio head flutters. In most VHS machines the upper drum is a tight fit on its shaft and no provision for centring adjustment is made or needed. In other cases, true running of the head can only be achieved with a suitable (and rigidly-mounted) dial gauge calibrated in μm. So long as the drum is centred to within 2μm, no trouble should be encountered.
4 Sticking of the tape to the drum surface, caused by condensation of moisture on the latter. This is prevented in all portable VCRs, and many homebase types, by a dew sensor 45 on the metalwork – more details in the next chapter. The patches of tape-coating on the head drum usually remain till scrubbed off, manually, and the afflicted tape (see Figure 12.11(f)) will trigger the end-sensors whenever the affected section comes out of the cassette.

guide. For the entry side check the perpendicularity of impedance roller, FE head and the first static guide; and the height setting of the supply reel disc 20. For the exit side check the perpendicularity of any impedance roller 41, the audio/control head stack 7, and the capstan 38. If these are in order, suspect the pinch roller 37 then the fixed guide immediately upstream of it, 42.

Apart from incorrect entry guide adjustment, a possible cause of 'bottle-necking' at the beginning of each head's sweep (top left waveforms in Figure 12.5) is low back-tension; the same effect at the end or centre of the sweep-envelope may be due to an eccentrically-mounted head if it's confined to *one* head's sweep.

When the tracking is perfectly correct, adjustment of the tracking control should result in an *even* rise and fall of the whole envelope pattern. Any tendency for the envelope to bottle-neck or develop a wedge-shape as the tracking control is turned indicates that the tape tracks and head sweeps are not parallel, suggesting that guide alignment is incorrect, or that there is dirt on the rabbet. Cleaning instructions will be given later.

Impedance roller and audio head stuck

On emergence from the exit guide the tape encounters (in some machines) another impedance roller 41. As before, this is a trouble-free component so long as it is perpendicular, free to turn, clean and has an undamaged surface. If jammed or rough-running on its shaft, it can cause 'wateryness' in the vertical features of the reproduced picture – as can any mechanical problem which imparts jitter to the tape movement/tension, or prevents smooth running of the head drum.

Next comes the audio head stack 7, which incorporates at its top the audio head, at its bottom the control track head, and in most cases an additional audio-only erase head just upstream of the audio head. So far as mechanical considerations of tape-running are concerned, this stack must be truly vertical to prevent the tape being biased up or down in its journey past. There are several other aspects of this head-stack: fault symptoms will be discussed in the audio and servo chapters, and adjustment in the section headed 'replacement of deck components' later in this chapter.

After the audio head the tape encounters a fixed guide 42, adjustable as to height. Its setting seldom needs adjusting because it rarely has to be replaced, and does not wear at a fast rate; and because the audio head is adjusted to conform with the tape path rather than vice versa. When adjustment of this guide *is* required it is carried out by reference to a gauging block for a fixed height above the deck surface; and confirmed correct by an absence of tape wrinkling or creasing at its flanges. If tape damage does occur at this (correctly-adjusted) guide, the likelihood is that the audio head face has excessive 'tilt-out' at top or bottom; or that the pinch roller 37 is faulty (see below).

Capstan, pinch roller and take-up

The tape is pulled around the deck by the action of the capstan. It is important to realize that the friction between the shiny capstan shaft and the tape is not sufficient to drive the latter, so that the motion is transferred from the capstan 38 to the pinch roller 37 and thence to the tape. For this system to work the pinch roller must have intimate contact with both capstan and tape: this can be upset by insufficient spring pressure on the roller (a pull of about 2.5kg should *just* lift the roller from the capstan); or the build-up of rings of dirt on the capstan shaft at points corresponding to the tape edges. Either can cause intermittent or jerky tape transport with resultant wow on sound and mis-tracking; or bring the tape to a standstill, whereupon the syscon section enters stop mode after a few moments.

The pinch roller and its mounting can contribute

other problems. If the roller is shiny it may slip; if it bears against the capstan 'out of square' the tape rides up or down in its passage, to be chewed at one edge on the flanges of preceding or succeeding tape guides – or, in VHS decks, the take-up spool. These 'tilt' effects can also come from a capstan shaft which is not at right-angles to the deck surface. Check with the manufacturer's gauging block, and expect any problem to be due to incorrect mounting of the direct-drive motor or (in belt-drive systems) of the capstan bearings. On rare occasions, the deck surface 1 itself may be buckled slightly as a result of the machine being dropped.

An egg-shaped pinch roller (though the defect is not visible on inspection) causes wow on sound, while worn bearings in either capstan or pinch roller can give rise to sound flutter or wow. In all cases of incorrect tape running in the region of the capstan; sound wow/flutter or level variation; or tape damage, the pinch roller is suspect. Replace it, with its mounting arm where applicable. Apart from routine cleaning, this item is not serviceable.

The tape issuing from the capstan is wound onto the take-up spool in the cassette whose level must be correct in relation to the deck surface to ensure that the tape does not catch as it enters the spool. As with the supply spool, height adjustment is made with fit-as-needed shims under the turntable 21.

Motive power for the take-up spool is crucial. It must be sufficient to take up the slack tape, but not so great that there is appreciable tension in the tape downstream of the capstan. Manufacturer's specifications for take-up torque vary between 80 and 200g/cm. If it is excessive, tape stretching takes place, possibly with damage to one tape edge. If TU torque is insufficient the risk is that a loop of tape will form at the spool entrance, piling up in the deck and finally invoking stop (perhaps with tape damage) via the syscon, which senses the situation after a few seconds from a tape slack sensor or from the cessation of reel-rotation-sensor (27) pulses. More often than not, this happens on an intermittent or sporadic basis, most likely with a long (3 or 4 hour) tape in use.

Take-up tension is measured with a hand-held torque gauge like that shown in Figure 12.6. The gauge is held gently so that the pointer and scale slowly revolve at the same speed. To get access to the take-up spool with the deck operating it is necessary to use a scrap cassette whose tape spools have been removed, together with the upper plastic windows; and whose end-sensor holes at each front

Figure 12.6 *Hand-held take-up torque gauge in use. To facilitate this test most machines require the use of an empty cassette shell, just visible here in the 'well' of the VCR. The cassette lid of this top-loading machine has been removed for access*

corner have been blocked up. For some front-loading machines, top obstructions prevent this, in which case the entire front-loading assembly can be removed, then the cassette lamp 40 or end-sensors masked. By taping or weighing down any cassette-in switch 31 or slack-sensor lever, the deck can be put into play mode with no cassette present.

In most VCRs the take-up drive comes via a slipping clutch, which is usually itself responsible for out-of-tolerance torque. Before condemning or attempting to adjust it, stop the take-up spool by hand, and in the few seconds available before shutdown, observe where the slippage is taking place. Sometimes a faulty drive belt or idler (which may be above or below the deck baseplate) slips before the clutch does so, immediately identifying the cause of insufficient or intermittent TU drive. *Excessive* friction is always caused by a binding clutch where one is present at all.

In some designs of VCR, TU torque is directly adjustable with an *electrical* pre-set control governing current in the reel motor. Before adjusting it, stop the reel by hand and check the drive linkage for slippage where relevant. In most cases the drive motor should be stalled by this means. Where no adjustment is provided, check the reel motor current to be correct (or excessive) then suspect the motor itself.

Unthreading

Another aspect of reel drive concerns the tape unthreading action, during which the slack must be hauled in by one or other of the spools – which one depends on format, design and vintage of the machine. Failure to do so leads to a large loop of tape hanging from the front flap of the cassette, which usually gets caught in some deck component as the cassette is removed, damaging the ribbon. Here the reel drive motor or the reel idler assembly 22 is almost invariably to blame, but check that the motor is being energized (syscon section) during the unthreading stage. Do not confuse the large tape loop arising from this fault with the smaller slack loop produced by braking faults at the end of play, FF or rewind functions (see below).

Fast transport and reel braking

In fast-forward and rewind modes the tape is quickly transferred from one reel to the other while (in most designs) unthreaded. No clutch is involved here: direct drive from the motor gives a torque of typically over 250g/cm, any diminishment of which is generally due to slippage in the reel drive idler 22 or any belt used for transmission. In cases of fast-transport faults, it is important to establish that the main brakes are fully off; and that the drive motor is *attempting* to turn the tape spool. If not, a syscon fault (see Chapter 13) is indicated. The deck shown in Figure 12.1 unusually has electrical brakes under the spool turntables.

Any tendency for the tape to slacken and bounce during FF and REW should direct attention to the soft brake on the undriven spool. Its gentle pressure must be sufficient to keep the tape under slight tension during these modes. At the end of REW and FF functions the tape reels are stopped by 'hard brakes' whose pressure and operating sequence are crucial if slackness in the stopped tape is to be avoided. The brake for the 'trailing' reel needs to come on fractionally before its fellow to ensure this, though the situation is avoided altogether in late designs of machines, whose syscon micro-processors are designed to go into a short forward or reverse 'shunt' phase to normalize tape tension after fast transport.

In some machines the reel brakes are operated by a solenoid, often a magnetically latching type whose position is changed by a single pulse of current. It can stick, due to a faulty solenoid or stiff linkage, to leave the brakes permanently on (reel motor labouring, no take-up, etc.) or off (tape spillage and damage). Reel main brakes are most commonly operated from a lever linked to the tape

111

loading system and driven by a mode-motor or by the capstan motor via a 'switching clutch'. In all cases braking problems are generally solved by cleaning or replacing their friction surfaces and check/adjustment of spring pressures as specified in the service manual.

Betamax and Video-8 variations

Virtually all the remarks made above for the VHS deck are equally applicable to machines of other formats. Though the *shape* of the tape path is different for these, the operating principle and sequence of components on the tape path is similar, as shown by Figure 12.7 and 12.8 (Betamax) and 12.9 (V-8).

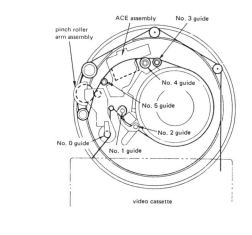

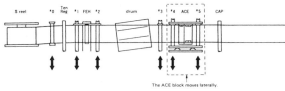

Figure 12.7 *Betamax tape path.* (a) *The original ACW loading system.* (b) *The resulting tape path*

For Betamax there are more guides than in VHS systems. After checking all other possible causes of tracking and tape running faults (see above) and thoroughly cleaning the tape path, guide adjustment is carried out as follows for anticlockwise-loading machines (Figure 12.7(a)). Screw up guides 1 and 2 so that the tape runs free past the full-erase head. Now adjust guide 0 upwards so that the tape ribbon

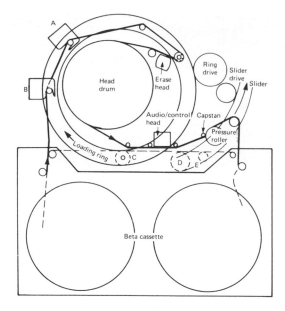

Figure 12.8 *Later form of Betamax tape loading system, with a smaller ring moving clockwise. This more compact arrangement renders the standard Beta path of Figure 12.7(b)*

just starts to lift from the rabbet lead-in on the lower drum. Now screw down guides 1 and 2 so that the whole length of their common upper flange plate is in contact with the tape, and biasing it downwards to correctly bear on the rabbet lead-in, as shown on an envelope waveform (Figure 12.5). Check that there is no wrinkling of the tape on guides or lower drum, and that it contacts the bottom flange of guide 0 and the upper flange of guides 1 and 2. Any problems here may be due to incorrect reel height or lack of perpendicularity of tension pole or full-erase head.

At exit from the head drum the tape contacts only the upper flanges of guides 3, 4 and 5. For adjustment all three guides are initially raised so that their flanges are clear of the tape, at which point the tape should rise clear of the rabbet lead out to give the 'crossover' effect shown in the top centre of Figure 12.5(b). With the tracking control offset to the right to give a reduced envelope amplitude, guides 4 and 5 (at the audio-head stack) are now screwed down to give a *rise* at the end of the envelope waveform. Finally adjust guide 4 for a flat envelope shape and check that this is maintained as envelope height varies with adjustment of the tracking control. If so, seal all the guides with paint or locking compound.

Late Beta machines have a smaller loading ring

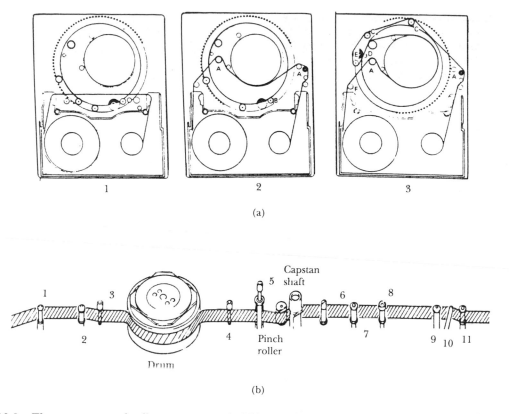

Figure 12.9 *The most common loading arrangement in Video-8 VCRs. In diagram (a) three successive loading stages are shown: 1 cassette in; 2 M-load phase complete; 3 end of U-load. Diagram (b) shows the Video-8 tape path in 'linear' form*

which rotates clockwise to give a tape path as shown in Figure 12.8. In essence the guide adjustment procedure is as given above, though some guides have provision for slant adjustment. Full details of path alignment are given in the appropriate service manuals. The pinch roller is most often responsible for poor tape running and tape damage in machines using this arrangement.

The effect on the screen of incorrect tracking is shown in Figure 12.10: at (a) entry mistracking, and at (b) exit problems. These correspond, respectively to the upper and lower waveforms shown on the left of Figure 12.5, and are relevant to all the helical-scan VCRs used in 'consumer' formats.

Most Betamax machines carry out FF and REW functions with the tape fully threaded. Slow and laboured 'fast' transport, especially rewind, in these machines is often due to excessive friction on the upper drum assembly, which here is a fixed cylinder below which the head disc rotates.

Replacement of this stationary upper drum is the only cure for this. Before doing so, check the reel idler and associated surfaces for slippage.

The ring-loading system in Beta machines gives little trouble. Sluggish loading action in machines where the capstan motor drives the loading ring may be due to belt slippage, but is more often caused by a faulty capstan motor; in designs where indentations in the periphery of the loading ring operate the reel brakes and pinch roller engagement by means of cams, a tendency for the ring to stick or jam is usually solved by lubricating the indentations and filing away any sharp edges they may have. Sometimes this type of ring-loading deck can emit a piercing squeak or squeal in operation, usually intermittently. It stems from one of the tape guide rollers on the loading ring being dry on its shaft, and can be silenced by a small drop of lubricant – beware of getting it onto the roller's outer surface where it will contaminate the tape.

Video 8 machines use a combination of arm- and

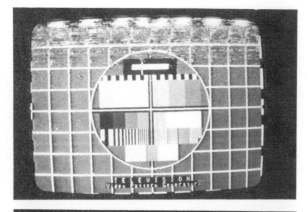

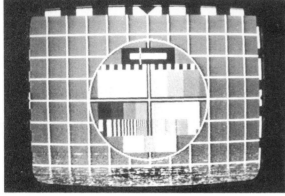

Figure 12.10 *On-screen effects of maladjusted tape guides. The mistracking shown at (a) is usually accompanied by picture rolling*

ring-loading to achieve tape threading to the tape path illustrated in Figure 12.9. Many of the guides shown here are purely to define the *shape* of the tape path rather than guide its running level. Guide 1 can be adjusted vertically and sets the running angle of the tape at the drum-wrap entry. Guide 2 is adjustable height and tilt-wise, and biases the tape onto the lower drum's rabbet to seat it firmly there throughout its traverse of the head drum; it is set for a flat envelope pattern at the entry side. Guide 3 merely absorbs tension fluctuations, like the impedance roller fitted to some VHS machines. The first guide downstream of the head wrap is no. 4, functioning primarily to damp tension fluctuations in the tape. Guide 5 can be adjusted for both height and tilt, and forms the main governor of the tape's exit angle from the head wrap; it is set for a flat envelope pattern at the right-hand (exit) side. Next comes the capstan and pinch roller, then guide 6, whose setting mainly concerns smooth tape running

in review mode. The purpose of the succeeding guides is merely to keep the tape clear of the head drum on its way back to the take-up reel; they are not adjustable.

Any slight tape-path errors in Video 8 decks are compensated for by the ATF (Automatic Track Following) feature of this format, which also tends to 'cancel out' any adjustments which may be made to guides: this necessitates, during tape-path setting, the use of a special jig (*track shift jig*) which releases the ATF system and substitutes manual control over tape tracking. Practical experience suggests that tape-path problems and mistracking are rare in Video 8 machines, such faults as do arise concerning tape crinkling, usually solved by adjustment of guides 5 or 6; or replacement of the pinch roller.

Alignment tape

For each format there is an alignment tape available from VCR manufacturers. Each is recorded on a special 'reference' machine at the factory, and contains recordings which, electrically and physically, are at centre tolerance for the format. Available on the tape(s) are colour bars; a monochrome step-wedge or test-pattern; and a sweep waveform, in which the carrier frequency is swept through all the frequencies normally encountered by the video heads – more details in Chapter 15. Along with these patterns, various standard sound frequencies and levels are recorded for test and adjustment purposes; these will be covered in Chapter 17. The monochrome portions of the alignment tape are used to give the envelope patterns used for tape path adjustment, e.g. Figure 12.5. Two golden rules are applicable to these expensive test tapes: *never* put them into a machine which has not first been *proved* mechanically safe; and use them as little as possible to minimize wear and the risk of damage.

For most purposes, it is not necessary to use the actual alignment tape. An acceptable substitute is an ordinary tape containing recordings of colour-bars, test card, monochrome patterns etc., along with one or more audio test tones. Make the recording on a new high-quality VCR which has first been checked to be on tracking centre with the alignment tape; then check in a second machine that the amplitude of the off-tape envelope signal is within 10% of that from the alignment tape. This substitute 'work tape' can be used for guide alignment, audio level checks etc, leaving only r.f.

sweep checks (see Chapter 15) and (in special cases) a final machine check for the precious alignment tape. It is not recommended to *adjust* tape guides etc. to any large degree with either of these reference tapes in the machine for fear of damaging them.

Cleaning VCR decks

Dirt, in our context, consists of tape debris and rubbed-off oxides and coatings; household dust; dried grease; settled deposits from the local atmosphere, e.g. tar, fat and salt; and many other things. They affect the operation of VCR decks in several ways, and we shall briefly examine each:

1 Formation of a barrier between heads and tape, impeding signal transfer. The symptoms of this are complete or partial loss of vision signals in a screen of snow or streaking. If one head only is affected, the picture flickers too, while in Video 8 machines the sound disappears along with the picture. Similarly the sound reproduction becomes low and muffled, or completely lost, if the audio head is blocked. Poor transfer of control-track signals due to a dirty CTL head causes loss of servo lock on replay. In some designs this mutes the sound and/or picture.
2 Creation of a *diversion* of a tape path. Dirt build up on crucial parts of the tape path (guides, head drum rabbet, etc.) can deflect the tape to cause mistracking or poor tape running.
3 Interference with mechanical functions. A heavy build up of dirt on the capstan shaft at points corresponding with the tape edges can 'stand-off' the pinch roller from the capstan and impair the roller's grip of the tape to give symptoms ranging from wow on sound to stoppage of the tape. Contamination of the take-up clutch leads to binding or slipping, and the latter applies to drive belts and idlers; all three of these categories are also subject to wear, with similar effects. Dirt on the tape-loading drive surfaces can prevent completion of loading, or heavily burden the loading motor. Dirt particles coming between the tape guide assemblies and their locating block V-notches 5 (VHS and Video 8) causes mistracking, shown on an oscillogram of the f.m. replay envelope shape.
4 Tape damage. Sharp or abrasive particles stuck on the tape path scratch or score the tape. If they

are present on a rotating member (guide, capstan, pinch roller) creasing of the tape results, with the visual effect of regular patterns of drop-out interference on the reproduced picture.

So a prelude to any deck servicing or fault-finding is a thorough clean of the deck surface and all the components in the tape path. Start by mopping up any dirt or debris on the deck surface with a cotton bud dipped in solvent. Now, using a cloth moistened with methylated spirit, isopropyl alcohol or surgical spirit, clean and polish all the tape-contact surfaces shown in Figure 12.3, 12.7(b) or 12.9(b) as appropriate. Give particular attention to the flanges of the tape guides and the rabbet on the lower drum, scrubbing the latter with a solvent-moist cotton bud to remove stubborn deposits. The capstan, too, often has a sticky build-up of dirt which is difficult to remove. Wet it with solvent initially, leaving the deposit to soften while other cleaning is in progress, then rub hard in a vertical direction with a soft cloth pinched tightly between finger and thumb, rotating the capstan continually. Finish by polishing the shaft surface. Clean the pinch roller and remove its shine with solvent and a soft cloth.

Head cleaning is left till last to avoid migration of dirt from elsewhere. Use a soft lint-free cloth soaked in solvent, stretched tightly across a fingertip to clean the video head chips and drum surface, stroking horizontally, (never vertically) to avoid damaging or breaking the fragile head tips. In some cases it is easier to hold the cloth stationary and rotate the head past them, always turning anti-clockwise. This fingertip method gives a useful indication of the state of wear of the head. With the moderate pressure required to clean the head, the tip projection can be felt (and, with experience, gauged) by its 'feel' through the cleaning cloth. An alternative cleaning material is buckskin; cleaning sticks tipped with this are available commercially. Cleaning of stationary audio and control-track heads is best done by scrubbing their surface with a solvent-soaked cotton bud, followed by polishing-off with a soft cloth or buckskin. In some Beta machines the surface of the audio-head stack is rather inaccessible – bend the cotton-bud stick and use a dental mirror (ideally an illuminated one) for checking and inspection. The same applies to some Beta and Video 8 capstans which are shrouded by a cylinder with side cutaway; for all formats and machine types the dental mirror is very useful for checking tape running over guides etc.

Deck service

At intervals depending on the amount of service it sees, and certainly when overhaul is required, the deck should be fully serviced. The main requirement of this is a thorough clean as described above, together with refurbishment or replacement of friction surfaces like idlers and drive belts; lubrication according to the manufacturers instructions; and a check of back-tension, take-up torque, brake pressure and tape running, adjusting or replacing components as necessary. Stretched or worn belts should be replaced; indeed, fitting a complete new kit of belts is recommended. Any rubber-tyred idler which shows signs of slippage should be replaced. Otherwise, degreasing with solvent followed by a light roughening-up of its friction surface with very fine glasspaper suffices. Never apply anything abrasive or sharp to the pinch roller, the drum assembly or any part of the tape path in attempts to clean and refurbish. A faulty pinch roller can only be dealt with by replacing it.

Lubrication must be carried out as the manufacturer specifies. Many bearings are not designed for external lubrication, and others, along with pinions, sliding surfaces etc., require a specific amount of a specific lubricant. Too much, or the wrong type, can ruin a component. Ensure that no lubricant finds its way onto friction surfaces or the tape path.

Tape damage

Tape chewing and ribbon damage of one kind or another is a common problem for the serviceman to sort out. Where the fault is permanent it is usually visibly obvious from a close inspection of the tape running around the deck. Sometimes tape damage takes place on an intermittent and sporadic basis, when it is difficult to be sure of the diagnosis or the certainty of the cure. To help in either case, Figure 12.11 shows various types of tape damage with suggestions for their causes. In many cases of tape chewing a vital clue can be gained by examining the deck surface for brown or black specks of tape debris. They will be thickest below the point on the tape path where the damage is occurring.

Figure 12.11(a) shows the effect of tape edge damage, where one side of the ribbon is wrinkled, scarred or serrated. It comes from binding against the lower drum, spool carrier or a guide, and usually upsets signal transfer of control pulses (bottom edge) or sound signals (top edge). Check for spool height, guide alignment, faulty pinch roller or excessive take-up torque.

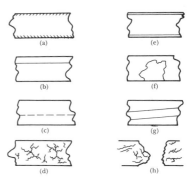

Figure 12.11 *Tape damage patterns: causes are described in the text*

Figure 12.11(b) shows a longitudinal crease in the tape, caused by a sharp abrasive particle on some stationary tape-path member – a fixed guide, the lower drum, FE head or audio head stack. Where the line (which may be at any height on the tape) is dotted (Figure 12.11(c)) the offender is on a *rotating* component: the capstan, a guide sleeve or the pinch roller in ascending order of dot- or dash-spacing. The crease in the tape gives a characteristic horizontal line of dropout on the playback picture as shown in Figure 12.12.

Figure 12.11(d) shows the scarring on the

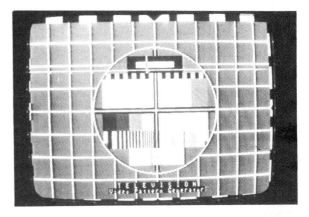

Figure 12.12 *TV-screen effect of a scratched tape, here causing a line of dropout about one-third down from screen top*

surface of the ribbon which can arise from the following causes: if regularly repeated along the tape, hairs or fluff are stuck to guide-sleeve, capstan or pinch roller; if just in one section of the tape, suspect a crunch-up due to failure of take-up during unloading (e.g. slipping reel idler) or formation of slack beyond the capstan – check take-up torque. If necessary, check also that the loading arms/guides 9 are fully retracted at 'stop' and that the cassette housing is locating properly. This type of scarring can also result from a loop of tape being crunched by the cassette flap during eject mode. Check reel-brake operation and unloading take-up torque.

Figure 12.11(e) shows the creasing at tape edges which can stem from incorrect height setting of guide rollers or guide pole, and Figure 12.11(f) the severe damage resulting from the tape sticking to the drum surface when an attempt is made to play or record with dew condensed on the drum surface – or too soon after head cleaning! The clear patch in the tape will be seen by the machine as leader tape, and the end-sensors triggered. The deposit on the head drum needs hard scrubbing to remove it. Some machines have dew sensors which prevent operation under these circumstances, and some of these (mains-powered) have a drum heater which is switched in to drive off moisture.

While still on the subject of the head drum, any *slanted* scars (at an angle of about 5° to the tape) must stem from an abrasive spot (e.g. chipped head ferrite) on the head drum itself, whose periphery runs at this angle to the tape, Figure 12.11(g).

Figure 12.11(h) represents the ultimate disaster, a cut or badly tangled tape. Actual breakage usually occurs when attempts are made to remove the tangled tape from the machine, or when it gets caught under the loading ring in Beta machines. This sort of foul-up is most often caused by failure of the machine to retract the tape into the cassette during play/record or unthreading. A rarer cause of tape 'severance' is where the ribbon end is torn from its anchorage in the reel due to failure of an end-sensor.

When a tape is damaged never put it back into a machine for fear of deck and head damage or (at the very least) blocking the video head with loose oxide 'scar' material. At the price of blank cassettes it is seldom worth attempting to repair a tape, though certain precious programme-material can be salvaged by the use of a tape splicing machine. Manual types are available at reasonable cost. Follow their operating instructions to the letter, and wear cotton

gloves whenever dealing with 'naked' videotape. Unless the splice is made to end-leader tape there will be several seconds of disturbance on picture during replay past the joint.

Replacement of deck components

Large deck components like head assemblies, drive motors and lower drum blocks are expensive, and in some cases time-consuming to fit and set up. In view of this, assess the economic viability of any such project before going ahead. A new direct-drive motor assembly may be fitted at great expense, only to find that the video heads are nearing end-of-life. A replacement head drum is difficult to justify fitting to a 7-year-old VCR for which other spares have been deleted, perhaps, and whose every mechanical part is worn. The lifespan of video equipment seems to be generally regarded as between 5 and 7 years.

Video head replacement

The most common replacement job on the deck is that of the video head assembly. Start by cleaning the entire tape path as described above. For standard VHS machines, unsolder the four leads on the top of the video head drum, remove two screws and pull the upper drum vertically upwards to free it. Unseal the pack of the new head drum and ensure that both the bottom of the new drum assembly and the top of its drive turntable are perfectly clean and free from dust and debris. Wearing cotton gloves and being careful not to touch the head chips, fit the replacement onto the shaft, ensuring that the colour-codes of leads and head terminals match. Strange symptoms like no-colour or no-replay come from mounting the head half a turn out. Resolder the leads, then where relevant (seldom in VHS) carry out a head eccentricity check and adjustment, details below. Now refit and tighten with equal pressure the fixing screws, ensuring that the holes in the drum assembly are concentric with the threaded holes in the drum turntable. This minimizes the need for subsequent adjustment of the audio/control head stack. Polish any fingermarks off the drum periphery.

For Beta machines, a preliminary to replacing the head disc is removal of the upper drum assembly, being careful not to lose any shims fitted to its periphery near its two mounting bolts. Fit the head disc, checking as above for correct colour coding and cleanliness. Many Beta (and some late VHS) heads require eccentricity adjustment. Screw down the disc fixing screws so that they are snug rather than tight, then fix a dial gauge to a rigid deck member (i.e. tape-guide pillar) using a jig (available from VCR maker) such that the gauge's feeler arm bears on the periphery of the drum at the point above the head chips. Be careful that the feeler does not touch the projecting head tips! Now turn the head drum by hand, whereupon the dial gauge needle will oscillate. Stop at the point of minimum deflection and tap the opposite side of the head disc to halve the total error reading, when there should be little movement in the dial-gauge pointer during subsequent head rotation. Repeat the operation until the peak-to-peak deflection of the pointer is less than $2\mu m$, then tighten the disc fixing screws, checking afterwards that the error is still within tolerance. Re-fit the stationary upper drum section, with its shims where applicable.

Video-8 head changing follows the general procedures outlined above, though there are more connections because of the flying-erase head. Multi-head drums of all formats are replaced in the same way – the difference here is in the subsequent setting-up.

Regardless of the format, several setting-up procedures are required after replacement of the video head drum. Do not play an alignment- or test-tape in a newly-repaired deck until it is proved safe; start with a less valuable tape. The first job is to check the tracking, and adjust guides if necessary as described earlier in this chapter. Also required is longitudinal adjustment of the control-track head, to be described shortly; record and replay head-switch point adjustment, covered in Chapter 14; record writing current set-up; and pre-amp response adjustment as outlined in Chapter 15. With the right equipment and some experience none of these take long to do, and reliability, good interchange and best performance are thus achieved.

Changing audio head stacks

Replacement of the audio/control track head stack 7 is possibly a more exacting task than that of video head replacement, though it is less often required. Bear in mind that in all respects the new head must be adjusted and set up to conform with the tape path rather than vice versa, so no *guide* adjustments

are involved here so long as their settings were correct originally.

Unsolder and remove the old head, making careful note of the wire connections and colour codes where relevant. Fit and connect the new assembly, and screw it down to approximately the correct height. With a scrap tape running (and for some Beta decks, using a dental mirror to view) adjust head height for equal spacing of the head-face insert above and below the ribbon as shown in Figure 12.13(a). This gives a roughly correct height setting.

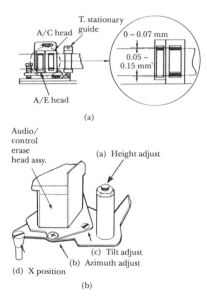

Figure 12.13 *Audio head alignment: (a) tape path past ACE stack; (b) common forms of head-position adjuster. The baseplate is pushed against cone d by a heavy spring, not shown here*

Play an alignment or test tape with audio test-tone, and adjust the head height (nut a in Figure 12.13(b)) for maximum output as seen on an oscilloscope hooked to the VCR's audio output socket. Now replay a high frequency tone (e.g. 6kHz on alignment tape) and with the scope connected as before, rock the head sideways with azimuth adjustment screw b to again maximize output level. Seal the screw at the best point, corresponding to a perfectly upright head. In a few cases a further adjustment may be required – to tilt the front of the head forward very slightly and ensure good signal transfer by slightly 'stretching' the tape across the audio (upper) section of the head. While this prevents audio drop-out and level variations, (check back-tension first) it must not be overdone. Excessive lean-out of the head's top edge may cause the tape to ride down and crunch on an adjacent tape guide; or may prevent good pulse transfer at the control-track head. This adjustment, called *zenith* or tilt, is carried out with screw c in Figure 12.13(b). From the starting point of a dead-vertical front head face, turn this screw no more than one turn to avoid trouble.

The final adjustment of the audio/control head stack concerns its longitudinal (X) adjustment: longitudinal, that is, from the point of view of the tape's control track. Sideways movement of the type of head shown in Figure 12.13(b) is made by adjusting the cone d for maximum f.m. replay envelope height from an alignment tape at the click-stop position of the tracking control. Some designs of audio head mounting do not have this cone-and-spring system; instead the slotted plate carrying the head is manually slid to the correct position while monitoring the f.m. envelope, then locked in place with its fixing screws.

Not all audio head assemblies have the single-nut height adjustment illustrated in Figure 12.13(b). An alternative (and more fiddly to adjust) arrangement has a three-point fixing in which all three screws together must be set for correct height, then individual ones adjusted as above for height and tilt settings. Any large adjustment of azimuth must be followed by a height check, then azimuth reset and so on until all aspects are correct.

Replacing other deck components

After fitting new tape guides, follow the advice given in the early part of this chapter in the three-stage adjustment sequence. The procedure for replacement of such items as direct-drive motors, lower drum assemblies and under-deck components depends entirely on the make and model of the machine, and varies widely – consult the setmaker's service manual, and ensure that the mechanism is correctly phased up afterwards. In general, it is necessary to set up the appropriate servo after capstan or head motor replacement; and to go through checks of tape path, head-drum servo and control-track head X-adjustment after replacing a lower drum assembly.

Cassette-loading mechanisms

For top-loading systems (now obsolescent in 'homebase' machines but much used in camcorders), troubles are usually confined to three sources:

1 Loss of tension in the V-shaped spring which operates the latch for the cassette's front flap – inner right-hand corner for VHS. Replace or re-tension it to ensure that the flap opens and rises as the cassette is lowered.
2 Insufficient pressure in, or breakage of, the spring-loaded 'fingers' which grip the cassette in the cradle, leading to floppiness and perhaps incorrect location. This can sometimes happen in front-loaders, too, to cause 'stuttering' and oscillation of the front-loading motor.
3 Failure of the damping mechanism, causing the cassette housing to fly up too quickly at eject. Replace the damper, which may be hydraulic, pneumatic or friction operated.

When re-fitting any cassette housing cradle, ensure that the holes in its bottom tray are concentric with the spool turntables, and that the tray itself is flat and not bent or damaged.

Front-loading (FL) mechanisms have more to go wrong, and are best removed from the deck for servicing. Put the FL mechanism on an insulating barrier (i.e. newspaper or magazine) alongside the deck while still connected electrically: this permits observation of the loading process. If required a second cassette can now be placed on the deck itself, and held down with a weight of about 500g on top, whereupon all normal deck functions are possible. To operate *without* the cassette, cover tape-lamp or end-sensors and hold down the cassette-down switch.

Check for freedom of the FL mechanism by turning the motor or drive pinion by hand, or by running the FL motor from a separate power supply, adjustable as to voltage. Ensure that the limit switches are clean and making good contact, and that the two sides of the mechanism are working in correct phase so that the carriage/cradle is parallel to the machine and deck front. Typical problems encountered in FL mechanisms are: 'bouncing' cassette and mechanism due to loss of tension or breakage in the cassette retaining fingers; cassette jamming behind entrance flap as a result of broken or misaligned sector gears on the flap levers; and mechanism overrunning (often with strain and damage) due to failure of limit switches.

Where the FL system is operated from a deck-mounted motor (capstan or 'mode'-motor) the main requirement during reassembly is to achieve correct phasing of the mechanism. The service manual specifies which mode (i.e., stop, eject) should be engaged while removing and replacing the FL assembly.

Excessive deck noise

Most video decks run reasonably quietly; the noisiest in normal play and record are those which use pinions rather than rubber idlers etc. for reel drive. Sometimes abnormal noise levels are generated, and in tracing them the important thing (unless the source is obvious) is to get as many clues as possible. Press the pause control: this stops the reels, guide sleeves, pinch roller and tape, and in many designs the capstan too. If the noise continues it comes from the head drum or its drive system, typically a bearing (rumble, squawk), or a belt or grounding brush (squeak).

If the noise stops with pause, analyse it (if rhythmic) in terms of observed revolution rate of pinch roller, spool, capstan, etc. A useful 'probe' is a pencil or similar sounding stick with one end applied to the ear and the other to suspect components and surfaces. Some typical noises and possible sources, based on practical experience, are as follows:

1 Squeal in FF and rewind modes: dry shaft at spool-turntable or counter relay pulley.
2 Squeal/squawk in modes involving capstan rotation: dry spot at bottom bearing of capstan shaft, or on pinch-roller bearing.
3 Intermittent squeal accompanied by picture tearing: dry sleeve-shafts on tape guides.
4 Noise from deck and on reproduced audio (longitudinal-sound machines): worn audio head vibrating.
5 'Bumping' noise on rewind (certain Beta VCRs): excessive tape friction on upper drum.
6 Roar/rattle during threading (some Beta VCRs): cracked pinion on loading pulley.
7 Scraping noise, depends on tape/spool speed: spool or turntable fouling cassette floor or cradle base. Check reel height, cradle latch etc.
8 Hum or roar from motor or other worn bearing: stop or slow the motion by hand to prove the culprit. Sideways pressure against a motor or bearing mount often modifies the sound to prove the culprit.

Motor problems

Electric motors used for drum and capstan drive are deal with in Chapter 14. The more humble motors used for reel drive, threading, FL drive etc. are less critical. To test them, spin the shaft by hand to make sure it is free-running, and rock the shaft to check the bearings. Turn the shaft very slowly by hand while measuring motor resistance with an ohmmeter. If the motor goes O/C at any point its commutator has a dead-spot and the motor should be replaced. Another problem encountered is build-up of carbon deposits between the commutator segments. This increases current consumption and reduces torque. Normally a motor can be tested on the current from the probes of an Avo or similar analogue multimeter on its lowest ohms range: the motor should self-start and run smoothly and slowly. Alternatively connect the motor to a source of about 3V d.c. and check that current does not exceed 20 or 30mA off-load. An oscilloscope connected across the motor running without a load should show a peak-to-peak ripple voltage of less than 500mV.

Worn brushgear or commutator causes sparking, which can give rise to interference (similar to static discharge) spots on the replay picture, or 'hash' on sound. To check for this, disconnect the suspect motor (in practice the reel motor) and watch for the disappearance of interference symptoms in the few moments before the deck shuts down.

Jitter

The replayed video signal contains timing jitter due to minute variations in tape tension and speed, head speed, etc. The degree of timing jitter depends mainly on the mechanical condition of the deck, motor bearings and so on, and is least in a new or newly serviced deck. The *effect* of timing jitter depends on the tolerance to it of the TV set's line flywheel sync circuit, whose time-constant, if too long, gives rise to a 'watery' effect in picture verticals, and often to picture-top bending or flagwaving. These effects are shown in the off-screen photo of Figure 12.14, and are worse from a second- or subsequent-generation tape copy,

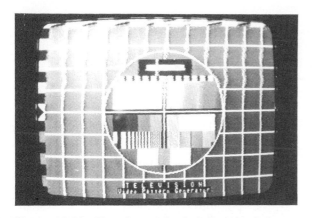

Figure 12.14 *Excessive mechanical jitter in deck components, or an incompatible TV set, causes this 'wateryness' of vertical picture features. Note the sideways pulling at picture top. This picture instability, though frozen in the photograph, is always moving.*

wherein the jitter from successive head-passes tends to add. Another effect of timing jitter is to upset the colour hue in the top section of the playback picture – it arises from poor phase-locking of the TV/monitor's subcarrier generator.

Some TVs have specially assigned channels or programme selectors for VCR and videodisc operation: these invoke a specially short line flywheel time constant for best results. Check the TV instruction book and ensure that they are used for VCR playback.

13

SYSTEM CONTROL IN VCRs

The system control (syscon for short) section of a VCR or camcorder commands and co-ordinates all the functions of the machine. In this respect it is similar to the TV control system described in Chapter 9, and the two have much in common. Except in very early (and now obsolescent) VCR designs the syscon is based on one or more microprocessors, which in some designs communicate between themselves via an I²C bus; or by some similar serial- or parallel-data system. The syscon section of a VCR is 'busier' than that of a TV set, and includes mechanical devices in its control loops and systems.

Between the many manufacturers, and over the years, a great diversity of syscon circuits have been used, ranging from simple to very complex. It is not our purpose here to attempt to explain their workings and principles of operation, other than from the standpoint of what goes wrong in practice, and where to look for the causes of specific symptoms. Very often, then, reference to the manufacturer's circuit description will be necessary if the key check points are not obvious from the circuit diagram.

Most problems which appear to be due to a system control malfunction actually stem from one of these: faults in the machine's mechanical section, as described in the previous chapter; faults, physical or electrical, in the several sensors dotted about the deck upon which the syscon depends for safety monitoring and status checking; or from problems in the in/out interface circuits. Only when all inputs are correct will the syscon permit initiation or continuance of a deck or other function. In syscon trouble-shooting a good rule is to suspect its electrical section (microprocessor and attendant chips etc.) last, initially treating it as a 'black box' whose outputs are assumed to be correct so long as its input requirements are met. If input conditions are correct, it's worth checking that the syscon outputs are reaching their destinations and there being correctly acted upon before delving into the central processing section itself.

Remote control is very much a part of the syscon section, and two types may be encountered: the now-rare wired type, which will be dealt with here, and the ubiquitous infra-red cordless system, covered in Chapter 9. In a VCR, IR command decoding is often incorporated within the main syscon chip, and as in TV sets, the main steps in diagnosing loss of remote control are to first ensure that the handset is emitting, and then to check that the pulse train is being applied to the decoder IC, whatever form it may take.

Following our plan of dwelling most on the areas which give most trouble, we shall start with an account of the sensors on the tape deck, examining their functions and the symptoms which arise when they trigger, legitimately or otherwise.

Deck sensors

The approximate positions of the deck-sensors in a typical VHS machine are shown in Figure 13.1; their siting will not necessarily be exactly as shown here, nor will they all necessarily be present – few *modern* VCRs have the slack sensor shown, and in

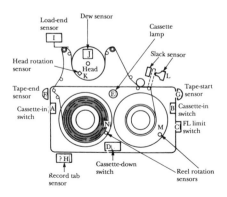

Figure 13.1 *Typical positioning of deck sensors in a front-loading VHS VCR. Design varies between makes and models*

many machines only the take-up spool carrier is fitted with a rotation sensor.

In a front-loading VCR the first operation is insertion of a cassette, which closes one or two cassette-in switches in the cradle, A and B in Figure 13.1. So long as power is available this should initiate rotation of the front-loading (FL) motor to draw the cassette in and down – if not check the switch(es), then for power to the syscon IC and to the motor. When the cassette is fully down the FL motor is switched off by a limit-switch C, generally mounted on the cradle itself; or sometimes by the action of the cassette body interrupting a light path between the cassette lamp E and the tape end-sensors F and G: in these machines leaky end-sensors are often responsible for front-loading faults which appear to be mechanical in nature. Checking the end sensors will be described shortly. Whatever the form of 'limit switch', its failure will generally result in the cassette being ejected a few seconds after its insertion.

The 'limit-switch' for top-loading machines is the cassette-down switch D, which closes when a cassette is pushed home. Failure of this switch inhibits all deck functions except eject: in any machine not fitted with a filament-type cassette lamp, this point should be the first check for a 'no-deck-functions' fault.

With the cassette in, the machine is ready to go. Recording is inhibited if the safety tab lever H is able to move into the cassette body – or indeed if it or its switch is stuck, faulty or dirty. A faulty switch or lever can also lead to the dangerous situation of permitting erasure of the valuable material on a pre-recorded cassette. If the deck refuses to respond to any tape-motion key, the likelihood is that the cassette lamp E is O/C: this is very common with the filament bulbs used in early VHS decks.

On pressing the play key (VHS and later Beta machines) the tape should thread while the head drum runs up to speed. If threading does not take place, listen for the loading motor, whose running suggests a broken or slipping loading belt. If there's no response to the play key, check if possible with the remote control, and try the other operating keys to narrow down the field of search (see later).

At completion of the tape loading the following actions should occur almost simultaneously: loading motor stop, reel motor start driving the take-up spool, and tape drive begin by either the pinch roller closing to the capstan or the capstan motor turning on. All these actions are triggered by the *loading-end* sensor I, which may be above or below the deck in the form of a leaf- or micro-switch (early machines) or a multi-contract *mechaposition* switch (late machines). If the load-end switch fails to operate, the syscon will within a few moments initiate unloading of the tape and reversion to stop mode. Most commonly this syndrome has its origin in mechanical problems as described in Chapter 12.

If the loading end-switch *is* operating, the unload-and-stop process could be due to several causes, initially investigated by a careful inspection of the deck at the moment when threading is complete. Check that both head drum and tape are moving, and that the take-up spool is rotating. If not, check the servos, syscon outputs and mechanics as appropriate. If the various motions *are* correct, one of the sensors may be giving a false signal. Fault-finding charts are given later in this chapter.

The head-drum itself is monitored in two ways by the syscon. In most machines there is a *dew-sensor* J on or near the head to detect condensation and prevent any deck action until it has dried: a 'dew' lamp on the machine's front panel or LCD readout warns of the situation, and in some cases a head-heater is switched on to speed the drying process. The main syscon check on head drum K is for *rotation*, by checking its flip-flop (SW25) pulses. If they are missing when the head turns, check for head PG (tacho) pulses from the motor or flywheel, then investigate the operation of the flip-flop in the drum servo circuit.

Some VCRs have a slack sensor L near the pinch roller. Failure of the TU spool to rotate allows the tape to bulge into the gap in an optocoupler, switching the deck to stop. More often, tape slack sensing is carried out by the *reel rotation sensor* M, sitting below the TU spool turntable. Figure 13.2(a) and (b) show the two most common forms it takes, in each case depending on the regular interruption of an IR light path. The output from the photodiode section should be a regular on-off pulse at a frequency dependent on the spool speed. If it's low or absent, check that the emitter LED section is passing current, that the receiving photodiode/transistor has operating voltage and passes current when IR light strikes it; and for reflective systems (Figure 13.2(b)) that the segments on the underside of the turntable are bright and clean. In many machines the sensors can be checked with a scope or voltmeter while rotating the turntable by hand with the machine in stop mode. Check also, where relevant, the supply spool rotation sensor N in Figure 13.1. Sometimes the absence of output from rotation sensors is due to connection troubles – dry-joints or loose grounding screws on the sensor PCB, whose effect (deck shutdown) can be intermittent.

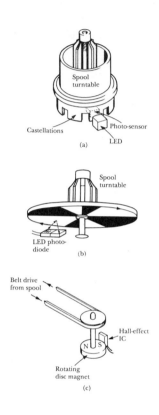

Figure 13.2 *Three forms of reel-rotation sensor, each described in the text*

An earlier form of rotation sensor is shown in Figure 13.2(c). Here the sensor is a Hall–Effect device within the alternating magnetic field of a rotating magnet, belt-driven from the take-up spool. The most common fault here is a broken or slipping belt, betrayed by a lack of movement in the tape counter where it is driven by the same belt.

If a deck-shutdown problem is due to lack of output from the rotation sensors, it can be proved by keying play, and then 'pause' as soon as tape loading is complete. Now the reel-sensor inputs are overridden, so the machine can remain in the play-pause mode without unthreading.

End-sensors

Mention has already been made of the tape end-sensing system in connection with front-loading machines. End-sensing is based on the action of cassette lamp E, which shines to left and right through the cassette body. With a fully rewound tape there is clear leader tape at the cassette's right-hand (TU) spool, permitting the light to fall on start-sensor photodiode G, whose resistance falls. with this LOW input, the syscon blocks any rewind command. Forward commands (play, FF, record) are accepted so long as no light falls on the end-sensor F. When it does, stop is invoked, followed in most machines by auto-rewind. This action may be triggered by any of the following:

1 The tape reaching its end and permitting the cassette lamp to shine onto the photosensor through clear leader tape (normal operation).
2 Electrical leakage in the sensor: in darkness the voltage across the sensor should be almost the same as supply voltage. Any less indicates a leaky photodiode/transistor. Replace it.
3 Light spillage through a damaged or chewed tape.
4 Extraneous light (bench- or table-lamp, sunlight etc.) reaching sensors.

Any of these things can cause the tape to go to stop (and usually auto-rewind) from play, FF and cue modes. The sudden termination of REW or review modes (to stop) can be due to the same effects on the start sensor (RH side of deck), but in fast transport modes, first check the counter memory system and the settings of any auto-index switches (VISS, VASS, QPF etc) which are often concealed behind doors or flaps.

In some VCR designs, continuous rotation of the reel motor (even without a cassette present) may occur due to an incorrect output from an end sensor.

In older VCR designs, the cassette lamp emits visible light from a tungsten filament. An incandescent lamp is easy to check, but unreliable, and its failure (O/C) causes the syscon to reject any command but cassette-eject. Later designs use LED emitters of invisible IR light: in general two are mounted back-to-back and connected in series. Again the cessation of current shuts down the deck, and here emission is not easy to detect. Check with an infra-red detector like those shown in Figure 9.2, or by voltage/current tests on the LEDs, but bear in mind that some portable VCRs and camcorders have a *pulsed* cassette LED system to save energy; and that some LED-current monitoring systems work on a scan/strobe principle in the same way as the key-scan circuit to be discussed later in this chapter.

Other formats

In general the Betamax and Video 8 formats have the same forms of deck sensors, as do the small-drum versions of VHS in camcorders and portable recorders. Some older Beta VCRs have a tape-slack detector in the form of a spring-loaded 'feeler' which bears against the tape on the LH side of the deck. Most versions of this are based on the action of a bar-magnet on the swinging lever, which moves over to operate a reed-switch when slack is present. Carefully adjust it to avoid false triggering or failure to detect slack. Back-tension or supply-reel problems can make it 'bounce' to intermittently shut down the deck. The lever can be simply taped back to override its action when investigating slack problems or during operation with no tape in the machine.

The main difference between Betamax and the other formats is in the method of tape-end sensing. The Beta tape has metallic leaders at each end, whose presence is sensed by a pair of coils corresponding to F and G in Figure 13.1. If oscillation in these coils stops, the syscon assumes that tape-start (RH coil) or tape-end (LH coil) is reached, and enters stop or rewind as appropriate. Where a Beta deck accepts only 'forward' commands, or only rewind command, check the RH and LH coils respectively: several volts peak-to-peak of sinusoidal oscillation should be seen across an operative coil. If not, check the coils themselves and associated oscillator components. Some old machines had resistive pre-sets to adjust sensor signal amplitude, which may be faulty or in need of adjustment. The same problems can of course lead to intermittent or sporadic shut-down or auto-rewind.

User interface

The messages from the deck-sensors described above represent only one set of input signals to the syscon section, albeit the most troublesome ones! A second set of inputs are the user commands, entering via the IR remote control section already described, or by the user's function keys mounted on the VCR's front panel. Very early machines used the operating keys themselves as levers to move deck members, and to close lever-type microswitches, the latter forming such 'command interfacing' as was necessary in these simple syscon designs.

Later machines use 'soft-touch' controls, with all mechanical deck functions directly controlled by the syscon via a 'mode motor', which may also play other roles. Two main interface systems are used between the user's keys and the central control unit, and will be described in turn.

Analogue–conversion system

The A-D conversion principle is used in, but not confined to, machines with full-function *corded* remote control systems. It is characterized by the presence in the circuit of a precision resistor ladder, represented by R1-R6 in Figure 13.3. In conjunction with R7 and R8, each button when pressed completes a potential divider to put a specific voltage into the inverting input of the comparator within D-A convertor chip IC1. The non-inverting input of the comparator is continuously fed with a multi-step staircase waveform derived from the four data line outputs from the main microprocessor chip IC2. With no keys pressed the comparator's output stays low at all times. When a key is pressed its characteristic voltage (typically ranging in steps from 10mV to 4V) appears at the comparator's

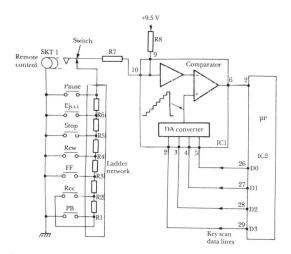

Figure 13.3   *A-to-D conversion system for VCR function control*

negative input. At some point in the staircase waveform at its positive input the comparator's inputs 'change over' and its output line goes high, indicating that a key has been pressed. The microprocessor now analyses the data on its output lines D0-D3 to ascertain the state at which the comparator output changes. This is compared within the micro with data held in internal ROM, a decoding process which results in the requested command being carried out. The advantages of this system are a minimum of connections to the user-controls; and that by duplicating the keypad and resistor ladder in a wired remote handset, all functions can be controlled via a 2-conductor lead. Its minijack connector plugs in at SKT1 in Figure 13.3. This jack is a trouble-spot: if its switch fails the front-panel keys are taken out of action, while poor plug contact results in permanent or intermittent failure of the remocon handset.

Apart from lack of operation due to jack-contact troubles, other faults can crop up in this type of circuit, generally taking the form of no response to the keys; or wrong responses, e.g. keying pause gives rewind, or keying play invokes eject. Where a wired remote control is available, it can be used for an immediate substitution test on the front-panel keys and their ladder network. In the absence of a remote control, disconnect the top of the network (i.e. R7 in Figure 13.3) and check for any leakage from its LH side to ground, which betrays leakage in one of the key-switches. This can happen for internal reasons, but check first for liquid spillage or other physical faults on the PC board.

If no leakage is present, check that the voltage at R7 starts high and drops to the correct characteristic voltage as each key is pressed; these voltages are usually quoted by the VCR manufacturer in the service manual or circuit description. If they are *all* incorrect suspect the ladder resistor; if one or two are wrong or varying, suspect the key-switches in question for high resistance; if one or more *individual* keys fail to work while the voltage at R7 remains high, the affected key-switches and their connections are suspect. The ladder resistor seldom fails. If the 'function-voltages' at R7 are correct (or if specific voltage information is not available!) check that the four data streams are passing into IC1 from ports D0-D3 on IC2. It is difficult to *analyse* these pulse trains, which appear quite random when viewed on a scope – if they are all present, though, the likelihood is that the problem is due to a faulty A-D convertor chip IC1: practical experience has shown that failure of this type of

chip is the most common cause of problems in such key-interface systems. If diagnosis-by-substitution has to be practised (often the case in this type of circuit) the A-D convertor chip is much cheaper than the other possible culprit, the main microprocessor IC2.

Key-scan system

An alternative key-interface arrangement is strobe-scanning of the user's function keys, used in TV and VCR IR remote control handsets as well as 'local' keyboards. Figure 13.4 shows the basic function. When pressed, each key bridges a key-scan line to a pulse-return line. The four (in this case) key-scan waveforms from IC output ports P00-P03 have different timings so that a strobe-detection process on any returning pulse at input ports PI0-PI3 indicates which key was pressed by reference to ROM in the drive microprocessor which generates the scan pulses and decodes the commands. This type of circuit is easier to trouble-shoot than the A-D system described earlier. Check that all keyscan lines carry pulses with different phases – if so the microprocessor is alive and probably well; if not, check the micro's operation, details later in this chapter. Now check that each key pressed passes the scan pulse back into the appropriate input port of the micro chip. If pulses are being returned with no keys pressed the likelihood is that one or more keys – or the PC board itself – is electrically leaky, perhaps due to corrosion, spillage or print faults (see Chapter 21). In making this test beware of return or output pulses which serve different *areas* of the machine by *function*-strobing. Refer to the manufacturer's circuit description if in doubt. So long as pulses are leaving and entering the microprocessor properly and no- or wrong-functions are taking place check the micro chip (details later).

In some equipment, especially portables, the key-scan circuit is extended to embrace such checkpoints as the deck sensors etc.

Some key-scan systems make use of a separate chip, mounted close to the keys, to generate the strobe pulses and decode the commands, then form them into a *serial* data stream for application to the main control microprocessor. This technique closely parallels that of an IR remote control system, but with transmissions via wire rather than a light beam; an example circuit is shown in Figure 13.5, where the similarity to a remocon system is

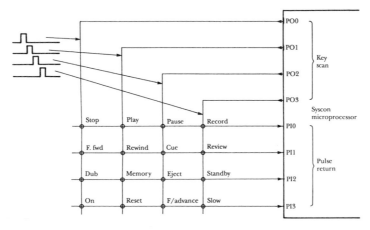

Figure 13.4 *Keyscan control system, basic operation*

seen in IC1. For troubleshooting, the same rules apply: check with a scope for key-scan and return pulses as before, then for the presence of serial data out, here at pin IC1 17. The (10-bit) serial data is only present while a key is actually depressed. Provided that operating power (IC1 pin 18) is available, and the crystal or ceramic oscillator (pins 2/3) is running, problems are most likely to be due to a faulty chip IC1; before condemning it for no output, check that the serial data line is not 'stuck' by a fault in logic inverters 31, pins 3 and 4.

Figure 13.5 also shows one form of microprocessor output expander (top and right) in which four output lines from the control micro operate eight indication LEDs via BCD-decimal convertor chip IC11. This and other forms of micro input/output expanders will be examined later in this chapter.

Peripheral syscon inputs

Depending on the type and purpose of the VCR/camcorder, there can be a third group of inputs to the syscon section; sometimes their function (or existence!) may not be obvious when they are the cause of a malfunction. Some of them are listed below:

1 *Timer input*: if the syscon thinks it is in timer mode (perhaps due to a closed timer switch concealed behind a flap or door) it will not permit the machine to switch on or function.

2 *Low-battery cutout*: in portable equipment the supply voltage is monitored, usually by an op-amp with pre-set voltage to one input (reference)

and sample voltage to the other. If the supply voltage falls below a pre-set level the deck goes to stop and switches off. This is misleading where shutdown occurs during a period of heavy current demand (i.e. tape loading) due to the power supply being unable to sustain the necessary current. Check the supply voltage with a d.c.-coupled oscilloscope, then if necessary the setting of the 'battery undercut' adjustment.

3 *Camera section*: portable VCRs have a pause line operated by a trigger on the camera. 'Stuck in pause' may be due to a malfunction on this line; in one case encountered the pause line was being energized by electrical conduction across the control PC panel surface due to leakage of electrolyte from a faulty timer back-up battery. Other 'separates' have a serial data bus for full remote control of VCR functions if required; checking procedure is similar to that for the I^2C bus (see Chapter 9), but consult the manufacturer's manual for details of the data format used.

In camcorders the camera and recorder sections are interlocked, command- and function-wise, by the syscon centre. It may not be possible to enter recording mode unless the camera part is up and running; similarly, some camcorders require the loading of a cassette, with record-safety tab intact, before the image-sensor and associated circuits are allowed to come into operation, typically by a 'record 9V' line governed by the syscon microprocessor under the control of (a) the safety-tab sensor; (b) the record key; (c) the cassette-down and (d) possibly the loading-end switch. Reference to the manufacturer's

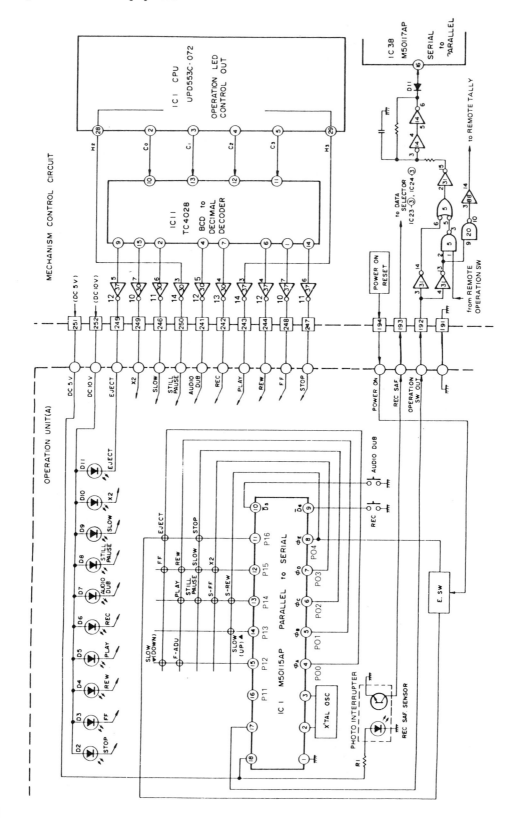

Figure 13.5 *Key-command conversion to serial data (left) and LED-drive decoding (JVC)*

service data is essential here to establish the existence and nature of the control links.

4 *Tape-position sensors*: every VCR has a form of counter-zero indication, ranging from a simple switch on a machine with mechanical counter to a 'high' indication from a comparator within a clock/counter chip. If this comes on, legitimately or otherwise, the deck goes to stop from fast transport modes. A second form of go-to-stop syscon input comes from the various tape-indexing systems VISS, VASS, QPF etc. Either a false input (disconnect to check) or the index-switch accidently left in the *on* position will invoke stop mode, seemingly at random.

5 *Power interrupt*: any momentary loss of power to the VCR results in immediate entry of the stop mode. Such things as a dry-joint in the PSU section, poor battery contact, a 'latchety' mains plug/socket connection or an intermittent fault on the microprocessor's reset line can invoke stop mode or complete switch-off of the VCR or camcorder.

6 *Remote control*: serial data enters the microprocessor or a peripheral chip for decoding into user commands. In some VCRs, high levels of noise on the IR input lines can trigger random functions in spite of the 'secure' encoding system used. It typically arises from a noisy IR receiving diode or preamplifier transistor/FET.

Some VCRs are also fitted up for remote control from an adjacent TV or other peripheral equipment via a rear-socket of DIN or SCART type; the incoming serial data has access to internal I²C, D²B or similar bus lines. If this or the IR control system is suspected for intermittent or random operation of the syscon, unplug or otherwise disconnect the data feed to prove the point.

Syscon fault-tracing charts

Figures 13.6 to 13.9 offer suggestions for fault-tracing the most common symptoms associated with system control. They are necessarily 'generalized' in that they cover, as far as possible, all types and vintages of VCR. Chart 1 addresses itself to 'light-touch' VCRs in which the on-off function is carried out via the syscon. For earlier machines operated by 'piano-key' levers, complete failure of the machine should lead to a check of the primary power supply (see Chapter 3) and particularly the discrete series regulators which produce various PSU output lines

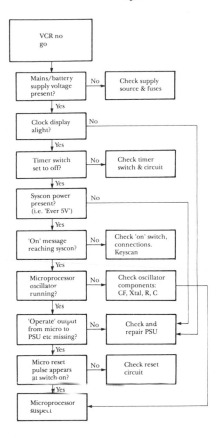

Figure 13.6 *Fault-finding process for a 'dead' VCR*

from the mains transformer/rectifier sets. In many 'electronic' VCRs the syscon microprocessor runs continuously from an 'ever' or 'always' 5V or 12V line, and brings the VCR to life by a 'high' output to one or more voltage regulators in the PSU section (see Figure 3.16), which may be in IC or discrete form. Failures of the various switched power lines (e.g. SW5V, SW9V etc.) is often due to failure of the PSU section to respond to the syscon's command.

Figure 13.7 is relevant to most VCRs; in the case of front loaders, failure to accept a proffered cassette should lead first to a check of the cassette-in switch(es), after which Figure 13.6 may be more relevant. For most Betamax and Video-8 machines, the tape is ring-loaded as soon as the cassette is inserted so here read 'failure to load tape' for 'no deck functions'.

Figure 13.8 is based on the fact that all syscon microprocessors are programmed to revert to stop after a certain time limit, during which the

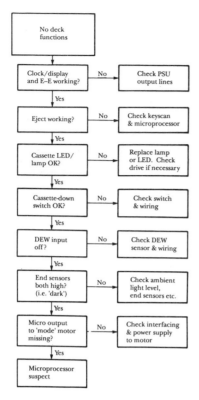

Figure 13.7 *Troubleshooting tree for a VCR with no 'mechanical' response*

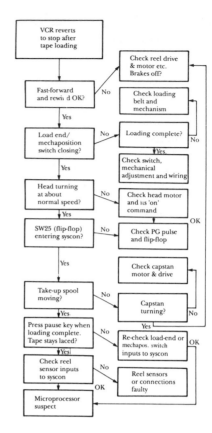

Figure 13.8 *A common problem is deck shutdown after completion of tape loading*

requested function (tape load/unload, reel drive etc.) should have been completed, as signalled to the syscon by the appropriate sensor: primarily the load-end/mechaposition switch and the reel sensors. These inputs are best checked at the syscon chip input pins themselves to prove the connections as well as the actual sensors.

Figure 13.9 covers the 'no rewind' function only, since virtually all problems (except failure of the key message to reach the syscon) relevant to FF function also affect other forward functions like play and record, and are largely covered in Figure 13.7. If both FF and rewind modes are inoperative, suspect the reel motor and its drive system where TU reel drive during play and record comes from elsewhere, e.g. capstan motor via a belt. In machines equipped with a 'mode' or 'mechaposition' switch, its adjustment, contacts and interface circuit is suspect for no FF or rewind functions. These often share a common mode-switch state/position.

Faults which only permit 'backward' modes, or which invoke sudden and unexpected entry into

auto-rewind have been covered earlier in this chapter in the section sub-titled *end-sensors*.

Syscon output interfaces

The syscon's output terminals are fewer than its inputs; where a function is missing or non-existent it is necessary to check that the interface is operating and the controlled device working before condemning the microprocessor system itself. Typical interfaces are shown in Figure 13.10: at (a) a two-transistor circuit operating a solenoid; at (b) a reel drive IC to control the speed and direction of a reel motor; at (c) two links to a PSU control chip for control of VCR operating lines; and at (d) and (e) direct lines to servo chips to control motor on/off switching.

The solenoid in Figure 13.10(a) may typically operate a brake- or pinch-roller, both of which would need to be off during play. If the solenoid fails to operate check that the micro's output goes

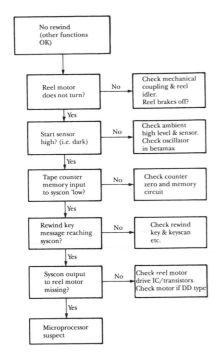

Figure 13.9 *Tackling a VCR which cannot rewind the tape*

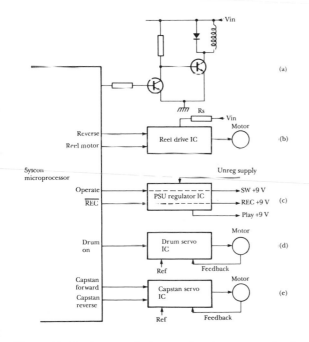

Figure 13.10 *Some typical output interface systems for the control micro*

low (SOL ON means low-to-operate) during the appropriate modes, then that the first transistor switches *off*, and the second one *on* to operate the solenoid, not forgetting that separate pull-in and hold arrangements are often present in the form of a double-winding or a charging capacitor (not shown here). If the pull-in section is not working, the solenoid will stay in when its plunger is pressed by hand. Some solenoids (especially in portable equipment) have hold-in magnets, so that just one current pulse in the appropriate direction is needed to toggle them. In case of trouble check the strength and duration (typically 200ms) of the operating pulse, then for excessive friction in the plunger and associated mechanism.

Loading and reel-drive motors may have two or three control outputs from the microprocessor. They go to an IC or multi-transistor buffer/bridge output stage. If the motor fails to operate, check *motor* supply voltage (i.e. safety fusible resistor Rs in Figure 13.10(b)) and the on/rev/forward outputs from the microprocessor, bearing in mind that a fault in the drive output stage can 'stick' the microprocessor outputs high or low. Disconnect IC pins to check if in doubt: motor, supply, and drive-circuit faults are more common than μp failures.

The power-line outputs from the syscon operate to switch the machine on, and to energize the record and playback sections of the signal-handling circuits. Thus the tuner, erase oscillator and recording f.m. modulator are powered during record; the f.m. demodulator, sound replay and chroma-up conversion circuits during replay; and the deck members, r.f. modulator etc. during both. Check that the syscon outputs are changing state correctly as appropriate, and again that a fault in buffer or invertor chip or transistor is not 'sticking' the output command line.

Figure 13.10(d) shows a 'drum on' command entering a servo IC, and Figure 13.10(e) the 'capstan forward/reverse' commands, either of which may in older designs be routed to a discrete transistor circuit rather than a chip. Failure of the deck motor to start up is rarely due to an absence of command from the micro: such faults as lack of motor operating voltage, faulty motor-drive amplifier or even a jammed shaft or flywheel is more likely. A meter or scope check on the syscon output will generally exonerate the micro itself.

Bear in mind that if a syscon output command is not carried out for any reason (electrical or mechanical) within a few seconds, shutdown to 'deck-stop' or 'VCR-off' mode takes place – it is programmed into the control microprocessor's 'emergency' routine.

Syscon heart

So far in this chapter we have regarded the central processor as a 'black box'; apart from the final boxes in the fault-finding charts we have concentrated purely on looking at input and output signals. When all inputs have been proved correct, and outputs are definitely missing or incorrect it's time to investigate the heart of the syscon itself. In simple machines there is a single microprocessor which takes care of all the functions described in this chapter, and sometimes the clock/timer system too. In these 'one-chip' syscons the three essential ingredients for correct microprocessor operation are (a) correct hash-free operating voltage, be it +5V, +9V or +12V; (b) that the system clock (generally internal to the chip, but using an external resonator in crystal, ceramic filter or CR form) is running; and (c) that a reset pulse is properly applied at power-on. Some microprocessors require a high pulse for this (pin marked RESET) while others need a low pulse (pin marked RESET). If these three conditions are satisfied, the microprocessor IC itself is suspect.

Sometimes a microprocessor can become 'locked out' by a surge on its supply line, due to mains power cuts or transients; lightning strikes; static electricity; or in portable gear, to reconnection of the battery without switching off. This lockout results in such symptoms as 'no-go', no deck functions, incorrect mechanical behaviour or no clock display. It can be overcome by completely disconnecting the machine from its power source for a period of seconds or minutes, then reconnecting power and switching on – a microprocessor reset is thus achieved.

In more sophisticated VCRs and camcorders a system of *port-expansion* is generally required to overcome the physical limitations of chip construction, mounting and wiring. One form of it is shown in Figure 13.11, where a special port-expander chip has sixteen input terminals, in each of which data (in or out) is latched until its address is generated on the three-bit address line at the right. At this point the data flows to or from the microprocessor, where it is internally held in RAM until updated by new data. In cases of trouble, check with a scope or logic probe for pulse activity on the address and data lines. The most common

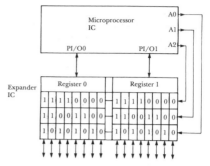

Figure 13.11 *Port-expansion chip. The two bi-directional microprocessor ports PI/00 and PI/01 can each handle eight in- or out-signals on a sequential basis defined by the data on address lines A0–A2*

cause of trouble is a 'stuck' line, generally due to a fault in the expander chip rather than the micro.

Broadly similar is the direct BCD-decimal convertor shown on the top right of Figure 13.5, and used for indication LED selection. Here four data lines C0–C3 give eight possible combinations at the output pins of IC11. Failure of a LED to come on, or several glowing together, is easily traced with a scope or meter: individual lamp problems are likely to be due to failure of IC11 or the invertors at its output; 'combination' troubles will stem from faults on data lines C0–C3.

Other expansion systems make use of two or even three microprocessors, communicating via parallel or (usually) serial data buses. One microprocessor is dominant, and it should be the subject of the first test once its identity has been established. In Figure 13.12 IC601 (CPU1) is dominant,

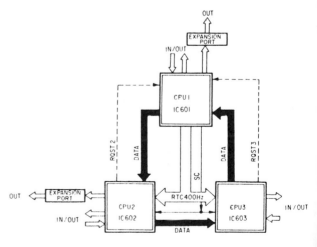

Figure 13.12 *Relationship between three control microprocessors (Sony)*

synchronizing the others by a 400Hz pulse RTC (Real Time Clock); 4-bit data is transferred in the sequence CPU1-CPU2-CPU3-CPU1, with each micro making its own decisions and producing its own outputs according to the data held in its ROM. Troubleshooting such a system is not easy, but can usually be achieved by checking for presence or absence of data (i.e. with a logic probe) on the data, address and clock lines – troubles are generally due to one being discontinuous, or being 'stuck' due to a faulty IC port. Analysis of the data is not possible with ordinary test equipment, though in practice diagnosis can generally be made without the need for complete pulse-train analysis.

Overriding system control

Sometimes it is necessary to override the syscon section for purposes of diagnosis. Simple methods of doing this include covering the cassette lamp or tape-end sensors: holding down a cassette-in switch or mechanical slack detector; shorting out a suspect dew detector; operating solenoids by hand; manually turning a spool to satisfy a rotation detector with no cassette present; disconnecting a suspect-leaky end sensor; and powering up mute and other lines from internal or external power lines.

A useful means of overriding system control and checking logic circuits generally is a logic pulser, described in Chapter 2. A signal fed from this (with print disconnected if necessary to check for a stuck line) should activate the logic operations of gates, counters and such syscon inputs as reel-rotation detectors. Applied to a ring counter such as that which drives a digital display or steps a LED indicator bank, the pulses will cycle the output through all states sequentially. A logic probe and pulser in use is shown in Figure 2.4.

14

SERVOS AND MOTORS

VCRs have two main servo systems: one for the head-drum and one for the capstan. In trouble-shooting these it is important to establish at the outset which servo is responsible for the problem; whether the trouble is basically mechanical or electronic in nature; and if the latter, which of the control loops – speed or phase – has got into trouble. In this chapter we shall cover very little of the well-trodden ground of theoretical circuit explanation, concentrating instead on tying symptoms to possible causes, fault-finding procedures, and understanding the basic control loops with their key test points and reference and feedback signal sources. In order to fault-trace in servo systems logically it is essential to know what each servo does, and the characteristic effects when failure occurs.

Because servos (by definition) consists of closed loops, minor errors and faults, as well as drift and ageing, are automatically compensated for. The failure of a servo implies that the loop is broken, and at the 'severed end', be it electrical or mechanical, a large corrective influence appears in an attempt to restore normality. This and the fact that (during playback) the TV screen itself serves as an excellent fault indicator make fault diagnosis less difficult than it may appear at first sight. Sound reproduction quality is an important clue in diagnosis of servo problems: all references to audio in this chapter assume conventional longitudinal sound systems, so switch Hi-Fi VCRs to standard sound when judging servo performance.

Let's examine the roles of the two servos. The job of the capstan is to maintain a steady tape speed at all times, so on both record and playback the motor speed must be closely regulated. Capstan servo faults generally affect the pitch of the reproduced sound, and this is the main means of telling them apart from drum servo faults; both affect picture tracking. The drum servo also has need to maintain a constant speed of 1500 r.p.m. for all formats, the exception being portable VHS machines using a 41mm head drum, which turns at 2250 r.p.m. Slightly different head scanning speeds

are required during 'trick' (still search, × 2, slo-mo, etc.) replay. In addition to its speed-control function the head servo must also closely maintain the *phasing* of the spinning head-drum: during record to ensure that the start of each head sweep coincides with the last lines of the TV field being recorded; and during replay to establish and hold a head-scanning path down the centre of each video track on tape. We shall enlarge on these soon. Before embarking on in-depth servo fault-finding, check that the control-track head is clean and correctly aligned; and that both capstan and head-drum are perfectly free to rotate. Each should spin for several seconds when flicked with a finger. Check also, where relevant, drive belts and other mechanics (e.g. tape spools) driven from the capstan motor. If these are in order, gather as many facts as possible about the fault before attempting a diagnosis: see whether sound *and* vision are affected, whether the trouble is present on record, play or both, using another VCR where relevant. Using the block diagram for the model in question, and scope and meter checks at key points the field of search can be greatly narrowed before the soldering iron is brought into play.

Head-drum servo

The head-drum motor is controlled by a dual-loop servo system as shown in Figure 14.1, which depicts the situation during record. Correct speed is maintained by the loop on the right, which samples the tone frequency coming from a generator (FG) inside the drum motor; the F-V convertor generates a voltage proportional to frequency (hence speed) and feeds it in inverted form via the MDA (Motor Drive Amplifier) to the motor. This loop holds the drum speed very close to 1500 r.p.m.

When the speed is correct it is necessary to *phase* the drum to incoming field sync, and this is the purpose of the loop on the left of the diagram. Here a phase detector compares the timing of incoming

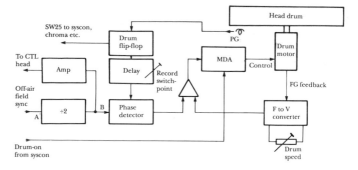

Figure 14.1 *Basic block diagram of a head servo in record mode*

field sync pulses with feedback from a PG (pulse generator) at the drum or flywheel, minutely controlling motor speed until the two are in step. Their phase relationship is set up by the record switchpoint pre-set, which governs a delay circuit in one or other of the pulse feeds to the phase detector. If this control is incorrectly set, the head-changeover point becomes visible on pictures as shown in the off-screen photo of Figure 14.2; if there is no phase control during record, this 'glass bar' effect drifts slowly up or down the reproduced picture, even when the tape is replayed in a good machine. So long as replay of a known-good tape is OK, this trouble is likely to be due to a lack of field sync input to the phase detector section. Check with a scope at points A and B in Figure 14.1, and bear in mind that the fault could be as far upstream as the vision receiver section, where poor l.f. response or a.g.c. problems can damage the field sync pulse.

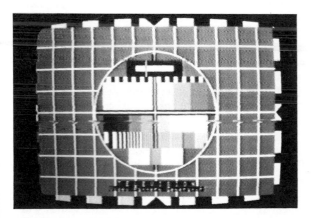

Figure 14.2 *The head switchpoint has here drifted to a position just below screen centre. It is over 10ms late*

A useful check point is the record pulse fed to the CTL head. If the symptom shown in Figure 14.2 appears by itself, concentrate the search on the head-drum phase loop. If it is accompanied by a cycling mistracking effect like that shown in Figure 14.8, the cause will be an absence of field sync input pulses to the servo section during record as described above. Under these circumstances of no control-track pulse recording some VCRs may mute sound and vision during replay (see later).

Speed faults

The problems described so far are confined to phase control of the head drum during record. Other drum servo problems generally manifest during playback too, and are easiest to track down during replay of a good tape. Figure 14.3 shows the basic servo operation during replay. It is very similar to the record set-up: the speed loop on the right is exactly as before to hold the drum at correct speed.

If the drum speed is far wrong the effect on the TV/monitor screen is loss of line hold as shown in Figure 14.4. The most common manifestation of this is very high drum speed (sometimes so high as to give no semblance of a picture) due to loss of the FG feedback signal; failure of the reference frequency at point A or B; malfunction in the F-V convertor; or a faulty MDA. In these circumstances slow the head down to about normal speed by pressing on its top with a finger. If there's no colour when the picture locks in, it may well be that the 4.43MHz crystal oscillator in the *chroma* section of the VCR has failed: this is often used in modern machines as the 'rock' reference, feeding point A in Figure 14.3. Where, as is often the case, the same

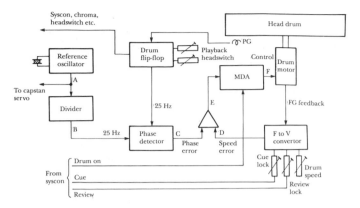

Figure 14.3 *Head servo in playback mode. Note the similarity to Figure 14.1*

Figure 14.4 *Loss of line hold due to wrong drum speed*

crystal paces the capstan servo, this fault also gives rise to incorrect tape speed and wrong sound pitch.

If slowing the head manually renders pictures in colour, the first check should be for FG input to the F-V convertor. If it's present try the effect of adjusting the drum free-run speed (alternatively called *drum discriminator* or *sample position*) pre-set, having noted its original position. If normal operation can be thus restored set up the preset according to the manual: so long as a small adjustment only is required the problem is probably solved, but if the pot has to be set near one end, or if resulting motor and MDA voltages are far from those quoted in the manual, the likelihood is that the circuit is faulty (more details later) or that the motor has developed a fault. The brush-type d.c. motors used in early models are most likely to be guilty of this. Some guidance on motor faults is given later in this chapter.

One way of accurately setting drum speed is to use a frequency counter to check line sync frequency from an alignment tape. Set for 15625Hz. In the absence of a counter, adjust for 'zero-beat' line-wise between tape replay and a broadcast transmission: various means of simultaneous display will suggest themselves to a perceptive technician, using a TV set or oscilloscope.

Phase faults

The other half of the dual-loop head servo is shown on the left-hand side of the diagram of Figure 14.3. The fundamental phasing requirement on playback is to set and hold correct tracking, whereby each head sweep scans down the centre of its appointed track. The basic reference signals for this are the control track on tape which shows the physical position of each video track; and the head PG pulse which indicates the physical position of the heads at any given moment. By tying these together in the servo phase control system correct tracking can be set and maintained. There are several ways to carry out this head-drum phase control; very often the capstan is used as an agent. A common system comprises phase-locking the PG pulses *and* the control-track pulses to a common master oscillator, via the head and capstan phase loops respectively. This master oscillator may be a dedicated type running at about 32kHz (e.g. Figure 14.3), or may be the 4.43MHz chroma oscillator as mentioned above. In either case the frequency and stability of the crystal also sets the basic speed of the motors on replay, ensuring that the output signals are close enough to broadcast standards to satisfy the TV or monitor.

The left-hand side of Figure 14.3 shows the drum phase-lock loop. PG pulses trigger a flip-flop whose pulse timing is compared to a divided-to-25Hz reference signal. The resulting error voltage is added to the speed control voltage and passed to the motor, whose phase becomes tied firmly to the reference oscillator. A phase error on replay will *not* be due to loss of drum PG pulses – their absence shuts the deck down via the syscon within a few seconds (see Chapter 13). In this servo configuration, head phase errors, with consequent mistracking, are usually due to the control-track processing loop, dealt with below. Problems in the phase detector or its feeds – or the frequency divider at point B – can lead to mistracking, however.

Drum instability

The drum speed determines the line frequency/ speed of the replayed video signal, so any short-term changes give rise to horizontal instability in the picture. Faulty motor or bearings, or jittering motor control voltage gives rise to horizontal oscillation or jitter of the image. To check whether the fault is electrical or mechanical, 'feel' the shaft when turning it by hand, checking for freeness and smoothness; then directly drive the drum motor at about the correct speed with an external source of very smooth d.c. voltage. If the symptoms remain, the problem is either mechanical in origin, or buried deep in the electrics of the drum motor.

Setting switchpoints

Figure 14.3 shows (centre) a pair of presets associated with the drum flip-flop bistable. While not directly concerned with the servo, their adjustment is generally regarded as part of the servo set-up process. They govern the point at which the replay head changeover takes place, and are set with a double-beam oscilloscope as shown in Figure 14.5: set each (Ch1/Ch2) so that its flip-flop transition takes place 6½ lines before field sync, which means approximately 4 lines before the start of field blanking, an easier point to identify in the waveform. The effect of incorrect adjustment here is a 'glass bar' (similar to that shown in Figure 14.2, but narrower) near the bottom of the screen. If the replay switchpoint is excessively late it may overlay

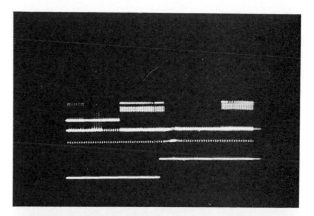

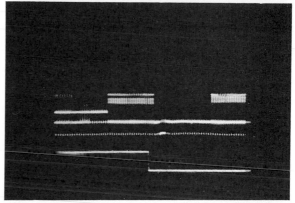

Figure 14.5 *Setting head-switch points. Field-rate oscillogram (a) was taken with the scope triggered on the negative slope of the SW25 squarewave below, and oscillogram (b) with positive slope triggering. Each is required in turn. The switching here is taking place about 2 lines (130μs) late*

and corrupt the field sync pulse to cause vertical judder and rolling.

Drum 'trick' speeds

In 'trick' replay, primarily cue and review modes, the drum speed must be altered by a few percent to keep playback line frequency correct. If line hold is lost (Figure 14.4) during these modes, check the operation of the drum speed loop switching. Very often separate speed presets are provided to trim for correct line speed in each mode, as shown at the bottom of Figure 14.3. A simple method of adjustment is to set for good chroma/luma registration on a known-good tape. Alternatively a frequency counter can be used to check correct output signal line frequency.

Capstan servo

Unlike the head-drum, the capstan has no 'absolute' phasing requirement, and is seldom provided with a PG system. All modern machines have a capstan FG built into the motor or the flywheel, and this provides feedback for the speed control loop, and in many cases the phase control loop too. Figure 14.6 shows a typical dual-loop capstan servo system of the type which may be used with the drum servo configuration described above. Again an F-V convertor is used in the speed loop, with a preset control for 'free-run' speed and a link from the syscon section to modify capstan speed for trick modes like picture search, slow-motion and pause.

The effect of incorrect capstan speed is a wrong-pitch sound track and a series of mistracking bars across the picture, which are usually unlocked (Figure 14.7). The number of bars is proportional to the speed error. If the sound is muted in the presence of this fault, check first the 'search' control line(s) from the syscon. As with the drum servo, capstan speed errors are most likely to be due to loss of FG or reference oscillator feeds to the servo section.

An accurate way to set capstan speed is to replay an alignment tape with a frequency counter hooked to the audio output socket of the machine: adjust the pre-set for exactly the audio frequency specified on the cassette's label.

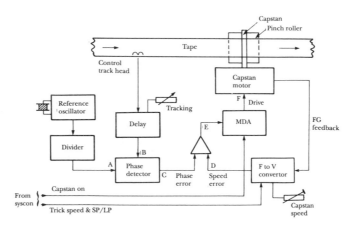

Figure 14.6 *Outline of capstan servo system, here shown operating in replay mode*

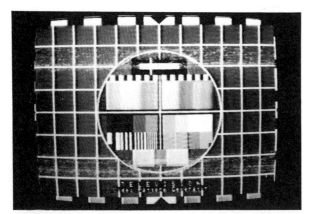

Figure 14.7 *Large capstan speed errors give rise to multiple noise bars across the screen*

Capstan phase faults

In circuits where tracking control is effected by the capstan servo (e.g. Figures 14.3 and 14.6) the loss of capstan phase control is most likely to be due to a lack of off-tape CTL pulses: if it only happens on the VCR's own recordings check whether the CTL pulse is being recorded. If the problem happens during playback of a known-good tape first ensure that the control track head is in good condition, clean and aligned correctly; and that the tape path past it is correct. So long as these points are in order, use an oscilloscope to check the two main inputs to the capstan phase detector at points A and B in Figure 14.6, and for the emergence of a phase error voltage at point C.

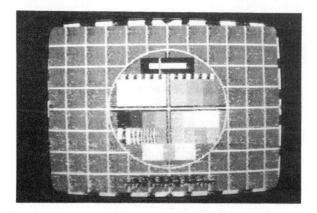

Figure 14.8 *Mistracking over the entire screen area*

Figure 14.8 shows the all-screen mistracking effect of incorrect capstan phase. If phase control has failed completely the effect cycles in and out of the picture, and can be controlled (and rendered almost stationary) by the capstan speed pre-set. A display in which the mistracking effect is constant suggests that the phase loop is working, but at the wrong point: try adjustment of the user tracking control. If it has an effect check the lateral setting of the CTL head (see Chapter 12 and the manufacturer's service manual) and the pre-set tracking control provided on some machines. If the tracking control has no effect concentrate on the delay block feeding point B in Figure 14.6; in some designs it operates in the reference path at point A. Phase faults in the capstan servo do not affect the sound to any noticeable degree unless the drive error is cyclic, when a 'wow' effect is audible, varying with the tracking errors visible on the picture.

Do not confuse the mistracking effect of a servo phase error with that resulting from permanent mechanical problems such as tape-guide misalignment. The latter are stationary, and generally affect only one area of the picture as shown in Figure 12.10 and 12.5.

All the fault symptoms described above for the capstan phase loop primarily concern replay mode. During record the requirement of the capstan servo is that it maintains a steady and correct tape speed in the face of changes in mechanical load. To achieve this, both speed and phase loops are used, with the speed loop being switched by the syscon between normal and half speed in LP-equipped machines; otherwise it remains in the form shown in Figure 14.6. The 'B' input to the phase loop is now switched to one of two possible sources: a divided-by-two incoming field sync pulse; or a pulse derived from the capstan motor's own FG source, counted down to a suitable rate.

Alternative servo configurations

Several different configurations for the phase loops are possible. The main variation on the one described above is to carry out tracking control on the drum servo during replay. If the replayed CTL pulses are also applied to the drum control circuit, no switching of the capstan servo is needed between normal record and playback modes. In servicing, consult the manufacturer's block diagram to establish how the loops are arranged, and check inputs and operation accordingly, always bearing in mind that the two essential timing inputs for replay tracking are the drum PG pulses and the off-tape CTL pulses.

Feedback generators

The main sources of feedback signals in VCR servos are PG, FG and CTL head. Problems in these generators and their connections are common, and checking them first can save much time. Pulse generators generally consist of a rotating magnet passing a stationary pickup head or Hall-effect device to generate a 25Hz or 50Hz pulse like that of Figure 14.9. Causes of little or no output are typically: the head and magnets are too far apart; the magnet becomes detached and lost; the head-coil or its connections are O/C (check also wiring, plug/sockets etc.); the flywheel or motor is

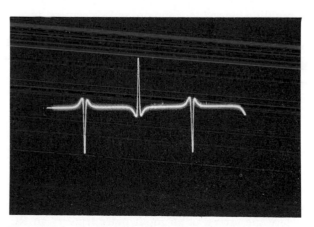

Figure 14.9 *Oscillogram of a magnetically-generated tacho pulse*

incorrectly assembled after service. Some pulse generators use a U-shaped optocoupler through whose gap a drum- or flywheel-mounted plastic or metal 'flag' passes. Check that the flag is not broken or missing, that the LED is passing current and that the photodiode is correctly biased.

Poor control-track performance is most often due to a dirty, worn or misaligned CTL head. In the first two cases the machine's replay of its own recordings are worst affected due to the signal passing twice through the faulty head interface. In many designs inadequate CTL replay operates a vision/sound muting circuit – override it by selecting play-pause mode.

Frequency-generators (FGs) are usually incorporated into direct-drive motors in the form of a printed or wound inductor over which a multi-pole magnet passes. Problems here are rare, and are generally due to open-circuit inductor or connections (check with an ohmmeter) or the rotor and stator sections not being close enough. On occasion the opposite fault may be encountered, where the circular magnet parts company with the rotor and binds on the tacho track. If caught in time (the first symptom is a rubbing noise) the magnet can be re-glued to the rotor; if the symptom goes unheeded the printed track is scored open-circuit, whereupon the FG signal is lost and the motor speeds up: this usually necessitates replacement of the motor. Other motor problems are discussed later in this chapter.

Servo electronics

The block diagrams presented so far have several 'building blocks' in common: pulse delayers, phase detectors and frequency-to-voltage convertors. It is not our purpose here to discuss these in depth, but for effective diagnosis it is essential to understand their basic operation, if only to identify the chip-peripheral components associated with each function – even analogue servos have for many years been buried in purpose-designed ICs.

Delay circuits

A pulse delay is provided by a *monostable* circuit, often abbreviated to *MMV* in service manuals. Here the arrival of an input pulse triggers an RC timing combination whose period terminates with an output pulse for use downstream in a phase detector: the delay period sets the phase relationship between reference and feedback pulses. In many cases the R element is adjustable (tracking control, head-switch preset, etc.) but the fixed C element is just as important in setting the time delay; it can usually be identified from the setmaker's IC block diagram.

Sample and hold

The essence of an analogue servo's phase detector section is a *sample and hold* circuit, consisting of a ramp generator and a sampling gate. The ramp is triggered by the arrival of one of the pulses to be time-compared. At some point on the ramp the other pulse arrives to open a gate and permit a sample of the ramp amplitude at that instant to pass into a reservoir capacitor whose average charge forms the error voltage, proportional to the time delay between the arrival of the two pulses. In many designs all three of these phase-detector waveforms – ramp, pulse and error output – can be seen with an oscilloscope on separate IC pins.

F-V convertor

The heart of the speed control loop is a circuit which provides an output voltage proportional to FG input frequency. This is also based on the action of a ramp generator – the amplitude of the ramp is sampled at its termination to form the output voltage. The F-V convertor is embodied in an IC or tiny module; if it malfunctions, check input waveform amplitude, fixed capacitors and the setting of the speed pre-set before replacing the chip or module.

Servo checking

Incorrect servo operation is always reflected in the relationship between the ramp and pulse waveforms shown in Figure 14.10. If the servo is out of lock the pulse runs up or down the ramp in the phase sample/hold circuit, but the *presence* of both ramp and pulse indicates that most of the phase correction circuit is OK. The likelihood here, then, is that the error voltage is not being developed or applied properly: check the sample reservoir capacitor and the route of the phase error voltage to the MDA. If either the ramp or the pulse is missing, check the

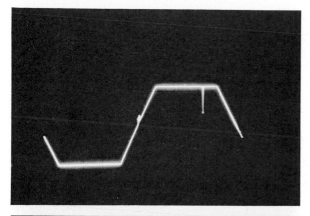

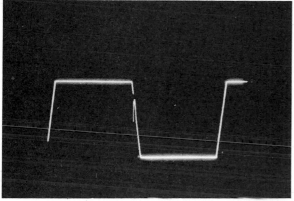

Figure 14.10 *Two examples of ramp-and-pulse oscillograms. The steep slope in (b) gives close, 'tight' control over the motor*

pulse inputs, ramp generator charging capacitor, etc. A fast movement of the pulse over the ramp indicates a speed error, directing attention to the speed control loop. The degree of jitter of the pulse on the ramp, and the amount of ripple on the error control line (s) to the motor are a good indication of the 'busy-ness' of the servo and thus the mechanical jitter and roughness of the motor and/or mechanics.

Figure 14.11 shows a typical IC-based servo system in block diagram form, in which most of the features and controls mentioned in this chapter can be recognized. The key points here are the master crystal oscillator at pins 1 and 28; the head PG pulse entering on pin 19; the incoming field sync (during record) to pin 8; and the drum motor control voltage (phase error output) leaving on pin 12. The capstan servo phase loop is also within the IC, depending on the master crystal again; the FG input at pin 22 during record; and the CTL track pulse entering on pin 4 during replay. The capstan

motor control voltage leaves the chip on pin 25. This type of block diagram is the best starting point for servo system fault finding, supplemented when the offending 'block' has been found, by the circuit diagram, which gives normal operating voltages and relates chip-peripheral devices to IC-internal functions. In this particular design the speed control loops are external to the chip, conforming to the right-hand sides of Figures 14.1, 14.3 and 14.6.

To get the composite waveform of Figure 14.10 from this particular circuit it is necessary to combine the ramp at IC pin 24 with the sample pulse at IC pin 27. A dual-beam oscilloscope, triggered from pin 27, can display both simultaneously; for use with a single-beam scope a simple R-C adder is connected between these test points. In other designs a test point is provided on which the ramp-and-pulse waveform is available directly. Consult the manufacturer's service manual.

When a servo fault is present it's sometimes difficult to decide whether the problem is due to the motor/mechanics or the servo electronics – and in the latter case whether the fault lies in the phase loop, speed loop or MDA. The easiest way of checking this out is to use an external power supply (which can be as simple as a battery and potentiometer) to inject test voltages at any point in the loops. Examples are the error outputs at pins 12 and 25 of the IC in Figure 14.10, corresponding to points C in Figures 14.3 and 14.6; the speed-error voltage points D in the latter two diagrams; and the input/output points of the MDA, E and F shown there. In general, correct speed corresponds to the voltages, given in the manual for normal operation, but beware of the situation (e.g. at one input to an op-amp) where a 'way-out' applied voltage is required to compensate for a fault or incorrect potential elsewhere in the circuit.

When injecting test voltages from a low-impedance source be careful not to overrun and damage semiconductors. If in doubt disconnect the normal 'upstream' source of the control voltage, or limit the current available from the external supply by use of a series resistor. Many mains-operated bench type PSUs have a built-in variable current limiter, which should be set to permit a few milliamps only to flow, except when directly feeding d.c. brush-type motors.

Whenever erratic servo operation appears to be caused by tape-tension or -friction problems, recheck the waveforms and error-voltage characteristics with an empty cassette shell installed: freed of the

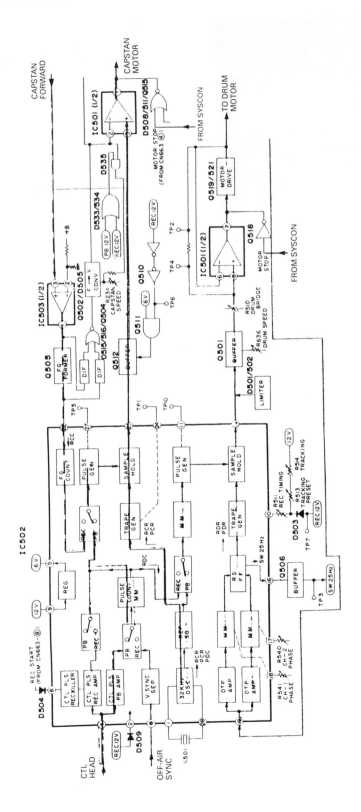

Figure 14.11 *Block diagram of commercial servo system (Hitachi)*

tape load the effect of motor, bearing, and friction faults can be seen.

Digital servo operation

A later type of servo system uses a pulse-counting technique for measuring time (phase loop) and frequency (speed loop from FG). From a fault-tracing point of view the IC may be treated as a 'black box' whose overall functions are the same as those given in the block diagrams earlier in this chapter. The counter-based chip gives very little trouble – indeed the wide adoption of this technique accounts for the virtual disappearance of true electrical servo problems from the technician's workload. All of our previously made points and diagnostic keys about reference and feedback sources and pulse-timings, speed and phase errors and symptoms, application of diagnostic 'steering' voltages and so on apply equally to this type of servo, so servicing is little different to before, especially as most problems spring from causes outside the actual servo-processing 'heart'. Digital servos have no pre-set controls of the types discussed earlier in this chapter: head-switching points, motor speed and phase control, and trick-replay compensation are determined with great precision purely by pulse-counting according to a ROM programme (fixed by the setmaker) within the servo and syscon chips. The main difference between this and previous systems, from the servicing point of view, is the way in which the error voltage is derived. As shown in Figure 14.12 the error/correction output from the digital servo chip is a PWM waveform whose mark/space ratio varies like the 'analogue' outputs of the TV control

chip shown in Figure 9.3. This varying duty-cycle can clearly be seen on an oscilloscope triggered from the rising edge of the PWM squarewave, and becomes a smooth d.c. control voltage after integration in the RC low-pass filter shown en route to the MDAs in Figure 14.12.

Fault-finding, then, devolves into checking and chasing the key inputs and outputs of the digital servo chip – these are outlined in the 'black-box' diagram of Figure 14.12. Some digital servo chips have separate PWM outputs and integrators for speed and phase loops, combined on their way to the MDA section. Pay particular regard to the inputs to the servo chip from the syscon, which are more comprehensive than with analogue servos: very often a 3 or more bit bus is used, or a serial data system using separate clock and data lines – reference to the maker's truth-table or serial-line data key is necessary to check that commands are legitimate and correct.

Video-8 servo system

Video-8 machines do not use a separate CTL track. Instead of 'field-marker' pulses along the edge of the tape ribbon, low frequency tones (100–163kHz) are written into the helical video tracks along with signal information. Each track carries a different tone from its immediate neighbours in a sequence of four, designed so that any tendency for the head to deviate from its correct track gives rise to a beat frequency of either 16kHz or 45kHz, depending on the direction of deviation of the head path. During replay the beat products are processed into a d.c. error voltage for corrective action in a perfectly conventional servo system. In troubleshooting

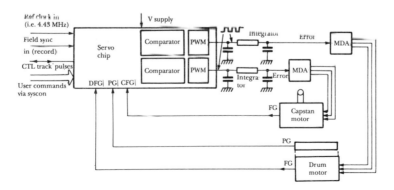

Figure 14.12   *Digital servo IC interfacing*

Video-8 servos, then, normal techniques as described above can be used for all faults except those involving the derivation of the phase error voltage. It should be emphasized at this point that a servo fault in a Video-8 format machine will not necessarily be in the ATF (Automatic Track Finding) circuit: indeed this is one of the most reliable sections of the whole servo system. Before getting involved deeply in ATF processing, then, check that the problem is not due to mechanical, physical or motor-related causes; and that the rest of the servo system (MDAs, speed loops, digital processor, system clock, PWM processing etc) is working correctly. Only when the symptom is demonstrably due to a faulty phase error voltage source should the ATF processing section be investigated; the main diagnostic tools are a double-beam oscilloscope and the manufacturer's block diagram and circuit description.

ATF processing

Figure 14.13 shows in much simplified form one method of processing the off-tape ATF tones, and indicates the key test-points to check during fault-finding on replay, which should of course be carried out with a known-good tape recorded elsewhere. The off-tape tones are separated from the other replay signals by a low-pass filter then a.g.c.-controlled to maintain constant level at point A. Here they enter a mixer, whose second input consists of a tone sequence similar to the off-tape one. It is locally derived in a programmable divider counting clock pulses from a crystal master oscillator working at 5.859MHz. The division ratio

changes once per field: over a four-field sequence the tones are the same as those coming off tape, though not necessarily in the same order. The presence and correctitude of the local tone sequence entering the mixer at point B is crucial to the whole operation: if they are missing check the low-pass filter and clock oscillator; if they are incorrect check that the programmable divider has its four essential instruction inputs, consisting of the SW25 drum flip-flop squarewave, SELECT 1, SELECT 2 and RESET inputs from the syscon section.

The beat products from the mixer consist of error indications at either 16 or 45kHz. Whenever one predominates a 'speed-up' or 'slow-down' output signal must be generated and passed to drum or capstan motor (usually the latter) to pull the tracking system back into phase. The beat products are separately intercepted by bandpass filters, whose outputs are individually measured in peak-detectors. The resulting d.c. potentials are applied to the inputs of a differential amplifier, whose output thus becomes 'steered' by the measured tracking error. Before this voltage is used for phase correction in the servo, however, it passes into a pair of sample and hold gates, triggered by lines TSA and TSB from the syscon section.

Sample/hold gate A is opened by pulse TSA, whose timing ensures that only during periods when the beat products represent valid tracking information is any notice taken of them by the servo system. The need for this arises from a deliberate timing offset in the locally generated tone sequence, during which pulse TSB (also sourced from the syscon) samples the error output

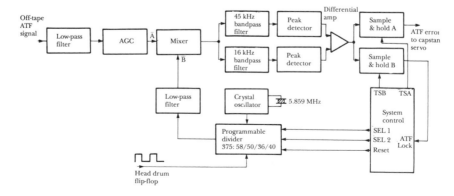

Figure 14.13 *Simplified block diagram of Video-8 ATF processing*

voltage. The output from sample/hold gate B is passed to the syscon in order to detect a false-phase-lock situation which can arise on starting or at edit points. The ATF lock output also senses, during replay, the speed at which the tape was recorded. Problems with playback SP/LP switching should start at this point, then.

Remember that with Video-8 systems there is no longitudinal sound track to give an indication of which servo system is in trouble!

Faults in any of the stages shown in Figure 14.13 cause phase errors in servo operation, and any fault which produces a d.c. error in the phase control line C gives rise to a speed error in the controlled motor. A suitable substitute voltage from an external source applied at point C reveals this by restoring correct speed. Any malfunction which does take place in the ATF process section is most likely to be due to lack of input signals (off-tape ATF tones, master crystal, SW25 pulse) or syscon commands (TS A/B, SEL 1/2, reset). Check also, where necessary, the 'physicals' of the filters (sometimes LC types) used for 16kHz and 45kHz beat selection.

Some of the components used in the replay ATF processing section are also used during record for laying down the tones in sequence, so useful clues can be had by making a recording on the faulty machine and replaying it on a good VCR. If playback is OK, the following parts are exonerated; clock crystal (here 5.859MHz); programmable divider; SEL and RESET lines from syscon; drum flip-flop pulse line continuity. This leaves the components in the upper section of Figure 14.13 to check for replay-only faults. By the same token, a VCR which replays correctly but whose recordings won't phase-lock the replay servo is most likely to be suffering from failure to record the ATF tones: check signal continuity from the programmable divider (tone source) to the recording heads in record mode.

Motors

Virtually all VCR capstan and drum motors devolve into two basic types: cylindrical ones with brushes, commutator and permanent magnetic field; and 'flat' types with multiple connections and Hall-effect coil switching via a special drive chip. The latter types are more expensive, but more reliable than brush types. Motor failures can lead to the following symptoms: lateral picture twitch, erratic line hold (drum motor); audio wow or flutter, loss of tracking control (capstan motor) mechanical noise, picture interference (either). The first checks on any suspect motor should be for freedom to turn, and worn bearings, both made by hand on the motor shaft.

Brush motors

Basically, brush motors are 'wearing' parts, especially the brush gear and commutator sections. Faults are betrayed by excessive ripple on a scope connected to its input leads: more than say 300mV pk-pk off-load for a capstan or head motor suggests the motor itself is faulty. Another test is to connect the motor, with no mechanical load, to the prods of an AVO8 or similar analogue multimeter, whose test current on the lowest resistance range should make it run slowly. If so, stall the motor many times by hand, ensuring that it restarts by itself each time. Still with the meter connected, turn the shaft very slowly by hand, looking for short- or open-circuit readings at any point. A momentary O/C reading indicates a 'dead-spot' which will sometimes prevent the motor from starting up; a momentary S/C or low resistance reading betrays the presence of conductive dust between the commutator segments. Either of these defects will cause rough and erratic running, and increase current consumption, possibly to the point where drive transistors overheat or a supply fuse blows.

Brush-motors with this problem can sometimes be dismantled and repaired by cleaning the commutator gaps with solvent, followed by polishing the commutator and brushes, then reassembly with the lightest touch of lubricant on the bearings. Even so, it's usually better to replace the motor, which after years of use will probable be in an advanced state of wear anyway. Set up the servo after motor replacement.

Occasionally brush-type motors develop sparking at the commutator to give ignition-like random interference on the playback picture – the effect is somewhat like that shown in Figure 15.5, though usually more 'concentrated'. The culprit can be identified by disconnecting the suspect motor to see if the effect clears in the moment before the picture breaks up (head motor) or becomes mis-tracked (capstan motor). Interference can arise in just the same way from a reel motor (see Chapter 12).

Hall-effect motors

Usually used in direct-drive (DD) roles, these motors have no moving electrical parts: the current in the multiple windings of the stator are switched by Hall-effect sensors via power transistors in a special drive chip. The rotor consists of a disc or cup-shaped magnet in which multipole magnetic fields have been 'printed' in manufacture. Apart from the FG-generator problems described earlier in this chapter, little seems to go wrong with these motors. The most common problem experienced by the author has been dried-up and seized bearings, mainly in the more exposed DD capstan motors, most of which are fairly easy to dismantle for cleaning and very sparing lubrication of the shaft and inner bearing surface. Sometimes this situation arises due to a missing or incorrectly-placed barrier washer at the bottom of the capstan shaft.

Such electrical faults as do develop usually involve one of the (typically six) coil sections, leading to overheating of the drive chip or fuse-blowing on occasion. If one coil-pair or -triplet fails, the result can be flutter on sound (capstan motor) or horizontal displacement of one section (usually about one-third) of the playback picture. This situation is revealed by monitoring the voltage (use an oscilloscope) across the low-value motor-current sampling resistor associated with the drive chip – identify it from the manufacturer's block and circuit diagrams. Imbalance between drive-coil currents shows as a ripple-voltage here. Unless you are sure that an electrical fault is due to the motor itself (and it's hard to be sure!) check the drive IC before condemning the motor. Replacement costs can be surprisingly high.

15

VCR LUMINANCE SECTION

A great advantage of dealing with faults in a VCR's luminance section is that (as with the analogue sections of TV sets) the monitor screen itself forms a very good 'fault read-out', so long as the symptoms displayed are carefully and thoughtfully analysed. Where a fault is present which affects the *basic* black-and-white picture, or its synchronization, the first essential is to establish whether it is present on replay, record or both – and whether the E-E (Electronics-to-Electronics, via the VCR's tuner in stop or record mode) picture is affected. The presence or absence of sound in each mode also gives vital clues – to the record/replay switching system; the operation of tuner/i.f. and r.f. modulator sections; and in Video 8 format to the head/tape interface system itself. If both vision *and* sound are missing in any mode, it is unlikely that the luminance processing block is faulty.

If signals are missing or impaired in E-E mode, check the tuner and i.f. sections as described in Chapter 7. If E-E *and* playback signals are missing, the r.f. modulator section of the VCR is the best starting-point for testing. Check its operating-voltage supply, whether a tuning-test signal is available, and whether signals are present at AV output terminals. Problems which are common to E-E and replay (i.e. 'flat' or negative picture, poor sync, etc) must be due to a fault in a 'common' path, i.e. the video output buffer or r.f. modulator.

First, then, establish just which section of the machine is in trouble by confirming correct E-E signal, then replaying a known-good tape in the machine. If it replays incorrectly or not all, diagnose the problem on replay before checking for faults on record. The exception to this rule, perhaps, is where head- and tape-interface problems are experienced: here it may be useful to make a test recording for playback on another (known-good) machine, whose picture reproduction can tell much about those sections of the machine which are common to both record and playback. More details later, and in the next chapter.

Initial fault-finding in VCR signal circuits is best done by reference to the block diagram in the manufacturer's service manual, which will in most cases be similar to the 'generalized' ones given later in this chapter, and give 'key points' for signal tracing and transfer. Once the offending stage has been identified, meter and oscilloscope tests will (hopefully!) pinpoint the individual components responsible for the symptom. This chapter will go through the entire luminance stages for record and replay, using many off-screen photographs to illustrate symptoms. We shall start (as is normal practice in an actual fault-finding situation) with the playback system, and in particular with what might be termed 'off-tape' faults, which account for the vast majority of replay picture impairments.

Luminance replay system

For fault tracing and setting-up purposes the luminance playback section may be conveniently divided into two parts: upstream ('source') and downstream ('processing') of the f.m. demodulator section. These handle f.m.-modulated and video-baseband signals respectively.

FM signal recovery

During replay the rotating heads scan the tape, each being switched into the replay circuit as it enters the tape wrap, and handing over to its fellow as it leaves, 180° and 20ms later. The f.m. carrier signal picked up is measured in μV and is the most fragile, delicate and corruptible form in which the video signal appears throughout the system. Problems in the f.m. recovery system manifest as noise and interference of one form or another, ranging from slight speckling or black-streaking to a complete snowstorm on screen, which obliterates the picture and resembles the display on a TV set tuned to a blank channel.

One of the most common causes is dirty or blocked

video heads. This generally arises from polluted or damaged tapes, and shows as a 'coarse-snow' effect on screen, as illustrated in Figure 15.1. At worst, no picture is visible at all beneath the snow; if only one head is dirty or blocked the snow effect has a pronounced flicker at 25Hz rate. Instructions for head cleaning were given in Chapter 12.

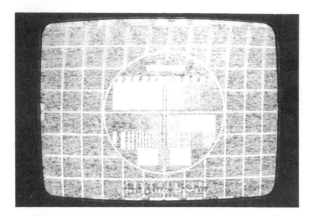

Figure 15.1   *Screen effect of dirty tape heads*

It may be that head cleaning has little or no effect on the picture reproduction, and the question then arises whether the heads themselves are faulty. Sometimes physical examination may show them to be damaged. If not, and if 'feeling' them by a finger through a thin cloth shows that they have reasonable penetration beyond the drum periphery, try making a recording and replaying it in a known-good VCR. If results are good, the problem lies in the f.m. pre-amplifier or associated circuits. Where (in 4-head machines) head-switching relays are used, these are particularly suspect. Slight contact resistance has more effect on the tiny replay signal than on the much-stronger f.m. record-drive signal; switching to and fro between standard and 'trick' (i.e. still) replay modes or tapping/flexing the relay body usually proves the point.

If (as is usually the case) the suspect VCR's recording replays badly in another VCR, the likelihood is that the video heads themselves are faulty. If possible check with a head-tester and/or use an ohmmeter to test continuity of the windings of the rotating transformers and heads; for VHS machines such checks can be short-cut by fitting a substitute head if available, since no setting up is required for a rough check of replay quality.

The same 'snowstorm' effect can come from *worn* video heads, though more often the symptom

here is a streaking effect from white verticals in the picture, and/or a 'raggedness' of the image, both of which are present in Figure 15.2. When the heads are worn, setting of the tracking control on replay becomes critical and the drop-out count increases. These effects can stem from causes other than worn heads, but the separate record and play tests outlined at the start of this chapter should eliminate them. Low f.m. output due to worn heads shows in an oscillogram of the combined f.m. head outputs (after the SW25 head switch), and the off-scope-screen photo of Figure 15.3 emphasizes the point by showing a situation where only one head of the two is producing low output; the lower trace in this photograph shows the SW25 head-switching waveform, which is also being used to trigger the scope's horizontal sweep. Test-point locations for these waveforms are given in manufacturer's service manuals for individual models.

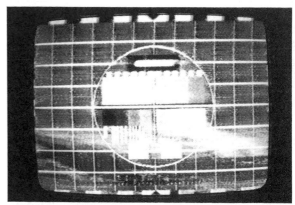

Figure 15.2   *Effects of worn video heads: dropout, black streaking, picture pulling, ragged verticals*

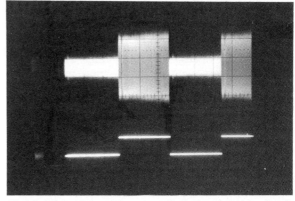

Figure 15.3   *Unequal outputs from the two video heads. The scope is triggered from the second trace (Y2) input, SW25*

The head-switching (SW25) system seldom gives trouble in a conventional VCR where only two heads are used at any one time for record or replay. Portable VHS machines with small head-drums have to use four heads progressively during record and playback, however, and this calls for a four-phase sequence of headswitching on both record and replay. If one head is blocked, worn or if one of the four head switches is inoperative, the effect is a 12.5Hz picture flicker which can only be sorted out with a dual-beam oscilloscope hooked to head switch (SW12.5) and replay f.m. envelope waveforms. Video-8 format machines use small head-drums in a two-head system; here head or interface problems result in loss of sound as well as vision – both are routed through the head interface.

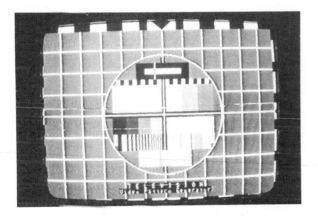

Figure 15.4 *Dropout. This image, recorded on a very worn tape, shows many dropouts, mainly in the lower half of the picture*

Dropout, illustrated in Figure 15.4, is the effect of momentary loss of replay f.m. signal. It arises from maladjustment or malfunction of the DOC (Drop-Out Compensator) section of the luminance replay circuit, though it can appear anyway if the tape in use is dirty or worn. Where a dropout compensator pre-set control is provided, it is normally adjusted so that the dropouts *just* disappear from a worn tape's playback picture. The first few minutes of a well-used tape contains enough dropout for this test. Maladjustment of a DOC control can completely lose the picture in a streaky-snowstorm effect.

Most cases of loss of vertical synchronization of a playback video picture are due to a defect in the f.m. envelope signal from the heads. The field sync pulses are recorded at the beginning of the head's

sweep of the tapes, so any shortfall in carrier level here corrupts the sync pulse train in the video output signal: see the waveforms in Figure 12.5(a). For a rolling or intermittently rolling replay picture, then, start by closely examining an oscillogram of the replay f.m. carrier signal, then refer to Chapter 12 if necessary.

Because the off-tape f.m. signal is so tiny it is vulnerable to micro-discharges of static electricity from the head drum mass. The effect of static discharge is shown in Figure 15.5, where white blobs can be seen appearing at random on the screen. It can usually be cured by attending to the head-drum grounding system, which generally consists of a contact brush on the drum shaft above or below deck. If there are many visible spots, and especially if they are in 'dash' rather than 'dot' form, interference from a motor on the deck is more likely to be responsible.

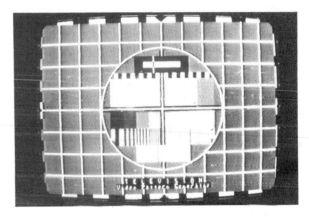

Figure 15.5 *The 'blips' on this picture are due to a build-up of static electricity in the head drum*

Figure 15.6 shows a block diagram of the pre-demodulator luminance replay circuits. The relay shown dotted is a feature of some machines only, and is used for switching between standard and 'trick replay' heads. More common is electronic head-pair selection, and record/replay changeover is invariably done electrically in transistor (very early) or IC (later and current) designs. These semiconductor switches can occasionally fail to give such symptoms as no replay, no record or one-head-only. Isolate the problem by studying f.m. waveforms in conjunction with switch control lines, and if necessary disable or enable the switch artificially for test purposes.

Each head has a separate low-noise preamplifier,

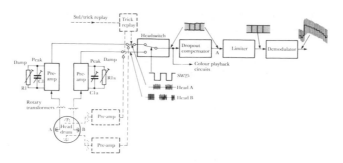

Figure 15.6 *F.M. luminance signal processing. Key waveforms are shown*

with two separate adjustments: peaking and damping, represented by R1/a and C1/a in Figure 15.6. These set the frequency response of each head/amplifier ensemble and are normally set up on replay of a special alignment tape containing a *sweep* r.f. signal. Each channel is set up for a specific waveshape on an oscilloscope hooked to point A in Figure 15.6. Generally peak response is set for 4.6–5MHz. Consult the service manual. A scope trace of a sweep waveform (correct adjustment) is shown in Figure 15.7(a). Since it is unlikely that both sets of adjustments would be incorrect (except, perhaps, after head-drum replacement) the effect of maladjustment or faults in the peak/damp or pre-amp section is generally accompanied by a flicker. Since f.m. carrier frequency is proportional to picture *brightness*, a likely effect of poor h.f. response (Figure 15.7(b)) in *one* head's channel would be a 'speckling' effect in the brightest picture features, flickering at 25Hz rate. Gross misalignment of head resonance presets can give picture effects very similar to those of worn or faulty heads.

Some VCRs (generally older ones) have a balance pre-set control with which the two heads' f.m. outputs can be equalized. Adjust it for equal outputs from both heads: if this control is wrongly set an oscillogram like that of Figure 15.3 is produced, and the monitor's picture begins to flicker when the f.m. carrier level falls below the clip level of the limiter which precedes the f.m. demodulator (Figure 15.6).

In still (freeze) and sometimes other trick replay modes the tape/head-sweep alignment is made such that unavoidable mistracking noise occurs during the field blanking period, out of sight of the viewer. This obliterates the field sync pulse, which is replaced by an artificial sync pulse generated within the VCR and timed by the head-drum flip-flop

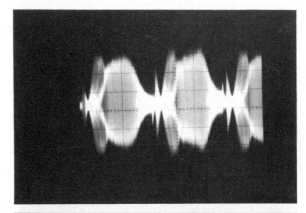

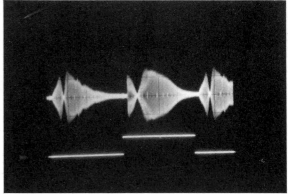

Figure 15.7 *Sweep waveforms from alignment tapes: (a) VHS; (b) Beta. The white 'pips' are frequency markers. Note the difference between the responses of the two heads in (b)*

(SW25) pulses. For the system to work a sample of replay f.m. signal is fed to a processing circuit associated with syscon and servo sections. Ascertain first that the machine is equipped for fine still reproduction before investigating noise-bar or field-sync troubles in this mode.

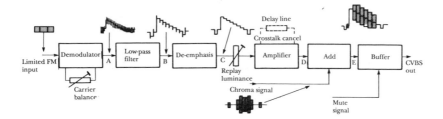

Figure 15.8 *Post-demodulator luminance signal processing. The carrier-balance pre-set is generally only found in older machines*

Replay luminance process

Figure 15.8 shows, in simple form, the luminance playback chain. As a general rule, problems upstream of the demodulator (dealt with above) introduce noise to the picture, whereas those downstream do not. No pictures with a clean, blank raster on replay will be due to a break in the signal line somewhere. Check with an oscilloscope during replay of a good tape, using the block diagram in the service manual, and bear in mind that this symptom is very likely to be due to the action, legitimate or otherwise, of a *mute* line. This can be overridden if necessary once obvious 'normal' causes of mute operation have been checked out. Typical sources of muting action are unlocked servos; and failure of the control-track head to pick up pulses from the tape, e.g. blank (unrecorded) tape or dirty CTL head. Most vision mute systems also take out the sound on replay, a useful clue to diagnosis. Where relevant, check the action of the machine's r.f. modulator and output buffer circuits by looking for a video signal at the AV output socket.

Many of the blocks shown in Figure 15.8 are embodied in ICs in modern machines – indeed some have but a single chip to perform all the functions shown. Even then, because these are analogue circuits many peripheral components are used, and the manufacturer's block diagram of the *chip* is sometimes required to relate these peripherals to the functions inside the chip.

The f.m. demodulator seldom gives trouble; when it does, noisy or negative pictures, or patterning of various types can result. The demodulator output voltage should be proportional to input frequency; any non-linearity causes the tonal values of the black-and-white picture to be upset, an effect similar to that of poor 'gamma'. The output from the demodulator is difficult to interpret on a scope

trace because it contains strong components at carrier frequencies. A recognizable video signal appears at point B in Figure 15.8, the low-pass filter output. A high level of carrier here causes beat patterning on the picture, and may be due to a faulty demodulator, incorrect adjustment of 'carrier leak' or 'limiter balance' presets, or a filter problem. Before condemning the latter, check that it has not come 'off-earth' for internal or bad-solder-joint reasons.

At point B in Figure 15.8 the reproduced stair-step waveform should be 'spikey' (due to pre-emphasis) but linear. Crushing of the top or bottom of the waveform gives a 'flat' reproduction of peak white objects in the picture (Figure 15.9) or poor synchronization respectively. Check voltages on the pins of the demodulator IC before changing it: supply-voltage problems or leaky capacitors can be responsible.

The crushed-white effect shown in the photo of

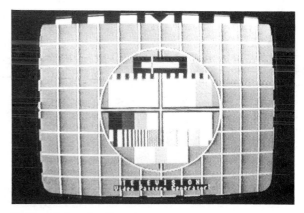

Figure 15.9 *Non-linearity in the luminance amplifier has 'crushed' the highlights in this picture, losing detail in the brighter parts of the picture*

151

Figure 15.9 can be caused by non-linearity (e.g. incorrect bias due to leaky coupling capacitor etc.) in any stage of the luminance replay chain. Check with a scope and stair-step or 'monoscope' test tape. In practice the r.f. modulator seems most often to be the culprit for this, especially where the vision symptom is accompanied by a buzz on sound: before getting involved with the modulator itself, though, ensure that its video input level is not excessive, and adjust the replay (and E-E?) video amplitude presets if necessary.

If the replay picture lacks contrast compared with the E-E picture, first check the condition and setting of the replay luminance level control, shown in the centre of Figure 15.8. If it needs a large adjustment, or 'runs out of road', a scope check at the key points between the blocks shown in the diagram will isolate the faulty stage – manufacturers are usually generous in the number of waveforms they publish for this section of the circuit!

The de-emphasis circuit follows the demodulator. It depends basically on a reactive (L, C) component for its operation. After de-emphasis (point C in Figure 15.8) the stair-step waveform should have reasonably square corners. If spikes are still present or if the luminance transitions have rounded shoulders (screen effects: 'sharp' and 'soft' pictures respectively) check the operation of the de-emphasizer. The same sorts of effects can come from a faulty or maladjusted picture-sharpener circuit, which uses some form of short signal delay to 'crispen' vertical picture transitions. Check the grounding connection of this delay line (and those of low-pass filter etc. used in the luminance replay chain) when the picture suffers from 'ringing' effects similar to those shown in Figure 6.2.

Poor definition of replayed pictures, where the tape is known to be good and the TV/monitor correctly tuned and in good condition, can stem from problems in almost any stage downstream from the f.m. demodulator. The trouble-spot can be pin-pointed by close examination with an oscilloscope of the waveform *corners* of a stair-step recording; or by looking at signal amplitudes or a 'multiburst' or frequency grating recording. Normally the response of VHS, Beta and Video 8 format systems falls sharply off at about the 2.7MHz mark; for S-VHS the figure is about 4.5MHz.

A part of the luminance replay circuit of many VCRs is a crosstalk-cancellation circuit, based on a one-line delay in glass or CCD form: it acts as a comb filter. The effect of luminance crosstalk on picture is similar to that of noise, but in a 'symmetrical' pattern looking somewhat like the pits on the surface of an orange.

Luminance recording

As with replay, many recording faults are due to problems in the head/tape interface area, so we shall concentrate here on faults which show up during replay of a tape recorded in the suspect machine and played back on a VCR (the same or another) whose replay capabilities have been proved good. It is always best to repair playback-mode faults first. They are usually easier to deal with, since the TV/monitor shows what's happening!

One exception to this general rule is the situation where the video heads are in an early stage of wear, when the machine may replay a test-tape satisfactorily, and its recordings play back well enough in another VCR; but replay pictures from its own recordings are noisy or 'ragged'. This arises from the video signal passing twice through an inefficient head system, and a temporary palliative can sometimes be found in (surprisingly) *reducing* the luminance writing current. Although this adjustment *should* be checked and set up when the head is eventually replaced, it might be wise, if recording current is deliberately reduced for this reason, to record the fact on a conspicuous note on the deck surface.

The luminance record section of the VCR is outlined in Figure 15.10. In fault-tracing, first establish that a correctly proportioned input signal is available at input point A: a broadcast test card off-air or a locally generated colour-bar signal applied to the video input socket is best. The first stage is for a.g.c. control of the signal, and the amplitude of its closely regulated output signal (point B) is crucial. Too low a level gives a flat, milky picture lacking in contrast; too high a level renders an over-contrasted image with patterning on highlights due to over-deviation of the f.m. modulator further downstream. The amplitude of the signal at point B should stay constant over a range of input levels: if not check the a.g.c. reservoir capacitor and for the presence of a line-rate keying pulse where relevant.

The next stage consists of a low-pass filter whose job it is to limit the frequency range of the luminance signal. Too wide a frequency range permits the f.m. modulator to deviate too fast, generating excessive sideband energy and causing patterning, particularly near sharp edges and fine

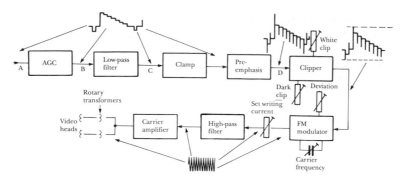

Figure 15.10 *Luminance record signal processing. In some designs carrier and deviation controls are associated with the clamp and a.g.c. blocks respectively*

detail in the image; the patterning may also affect the *colour* in those areas. There should be no vestige of chroma subcarrier signal in the output from the LPF, point C in the diagram.

As we have seen, the amplitude of the luminance signal governs the deviation of the recording modulator. Set up the deviation pre-set according to manufacturer's instructions, which generally involve the use of an accurate d.c. voltage supply and a frequency counter. The effects of incorrect deviation have already been described. Associated with the setting for deviation is that for *carrier frequency*, the modulator's 'starting point' with a sync-tip-level signal. If it is wrong the reproduced picture has incorrect contrast range and possibly wrong tonal values, especially in the lowlights (Figure 15.11). Again, patterning may also be present. Unless the modulator IC has been replaced, it is seldom necessary to disturb the carrier and deviation adjustments: check other possibilities

first, especially the luminance a.g.c. and *clamp* sections.

The next block encountered is the pre-emphasis stage, depending on a reactive component (usually capacitive) to give a 'lift' to high luminance frequencies. Few designs have provision for adjustment here; any malfunction generally shows as poor definition in the replayed picture, or as a 'fringing' effect on vertical picture features. The output from the pre-emphasizer (point D in the block diagram) has sharp spikes on transients in the waveform. To prevent these spikes causing momentary over-deviation of the f.m. modulator, white and dark clip controls are provided to limit the spike amplitude in each direction. Maladjustment of the white clip control leads to a black streaking effect from picture verticals (Figure 15.12) or a flattening of peak whites (see Figure 15.9); if the dark clip preset is over-advanced the sync pulses may be foreshortened, leading to jitter or roll of the picture.

The f.m. modulator is invariably inside an IC, and seldom gets into trouble. Check pin voltages and peripheral components before condemning the chip for no output, and (with the help of the manufacturer's IC block diagram) identify and check any small external capacitors associated with the oscillator in cases of incorrect output frequency with correct video (or d.c. test) input level.

The final stage in the luminance recording section is the carrier amplifier. It drives current (via the rotating transformers) through the recording heads, and this writing current is crucial to good picture reproduction. If it is set too high two effects are possible: a streaking from verticals similar to that shown in Figure 15.12; or a tearing/raggedness of the picture of the sort illustrated in Figure 15.2. There may also be some effect on chroma signal (see

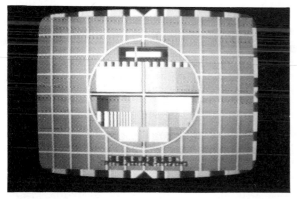

Figure 15.11 *This flat and slightly smeary picture was recorded in a VCR whose f.m. modulator pre-sets were maladjusted*

153

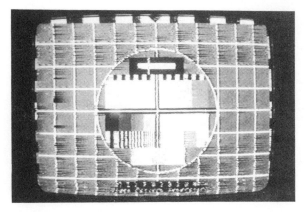

Figure 15.12 *'Over-deviation' noise in the form of black streaks from picture highlights*

Chapter 16). Adjust the f.m. record level (writing current) preset for the manufacturer's prescribed drive level.

A strange recording fault results from incorrect positioning of the audio and control-track head. If it is mounted too high, audio and servo operation may not be affected, but the control-track record pulses have the effect of erasing the first section of each (or alternate) video tracks. During replay, vertical sync is impaired, and an oscillogram of the f.m. luminance signal from the video heads shows a 'clean bite' at 20ms or 40ms intervals, see Figure 15.13. This effect disappears a second or two before the termination of the recording due to the spacing between the head-drum and the audio/control head assembly.

Failure of the machine to record a picture at all can usually be quickly traced with an oscilloscope

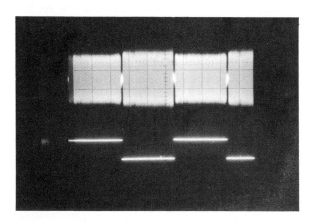

Figure 15.13 *Chopped f.m. carrier due to spot-erasure by a wrongly-positioned control track head*

and use of the manufacturer's block diagram. It is not a common fault.

Digital picture processing

Some VCRs incorporate a digital storage section in the form of a RAM memory with a capacity of typically 1–2Mbyte. A complete field of TV information can be stored here, and read out to form a noise-free still image, or to give a range of special effects. Where the 'trick' effects occupy the entire screen area, the CVBS signal can be sampled as an entity, known as *composite coding*. As shown in the simple block diagram of Figure 15.14(a), the video signal is sampled at $3 \times$ fsc rate (13.3MHz) to six- or eight-bit resolution. The serial data from the A-D convertor is transformed to parallel data for processing and application to the memory chip(s); it travels at fast rate on six or eight data lines to the memory chips in synchronism with the generation of memory address data. During readout, memory address data is once again generated in rapid sequence, and the binary data flowing from the memory banks is reconverted to serial form for application to a D-A convertor, whose output forms the analogue video signal. 'Geometrical' effects are generated by manipulating addresses, and other effects by reducing the bit-count of the digital signal or the latching rate of the D-A or A-D convertors.

In VCRs and TVs where the trick effects are 'mixed' with pictures from another source (i.e. a picture-in-picture display) it is necessary to use *component coding*, in which the video signal to be processed is split into three parts: luminance (Y), U and V, each for separate storage in 'parallel' memory banks. This is necessary to facilitate re-coding the chroma signal in accordance with the subcarrier phase of the main picture, whose burst signal it must share, and whose hues it must agree with. The general outline of this type of digital processing system is shown in the block diagram of Figure 15.14(b).

There is wide variation in the design and arrangement of these digital processing circuits; this and their good reliability record make it difficult to give specific advice on fault-finding, which should initially be based on the manufacturer's block diagram and the use of a double-beam oscilloscope. First establish that the ICs have correct operating voltage, and that the system clocks are working at correct frequency, based on a

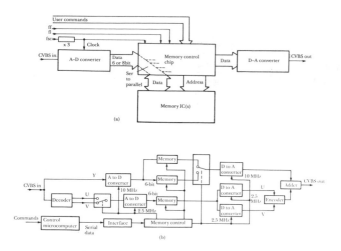

Figure 15.14 *In a field-store memory the video signal is digitized before being written into store, (a) for picture-in-picture, wipe and superimpose effects the chroma components must be decoded before being stored alongside the Y information in digital form. Here U and V samples are taken and memorized sequentially. Pal encoding of the read-out chroma signal is controlled by the 'main' picture's subcarrier timing (b)*

'chip-clock' (crystal or ceramic oscillator) and the 4.43MHz subcarrier reference fsc, which via a frequency multiplier forms the basis of the sampling and A-D conversion clock. If these 'heartbeat' pulse trains are present and correct, check that the system decoder- or microprocessor-chip is correctly receiving control data from the remocon and syscon interfaces, then concentrate on the LSI chip usually called the memory controller. In conjunction with the setmaker's block- and circuit-diagrams, check for the presence of data on each individual line in the control, address and data buses. As with teletext decoders, most problems are caused by 'stuck' lines due to chip, print or occasionally discrete-component faults. Picture problems which always show at the same point on the display are likely to be due to memory-chip faults. In picture-in-picture displays, incorrect hues, lack of colour, etc. are most likely to arise from faults in the analogue coder section, whose fault-checking pro-

cesses are similar to those given in Chapters 6 (decoder) and 18 (encoder).

Tape copying

The video signal is inevitably degraded during copying and editing. The effects may be a loss of definition; excessive edge emphasis; a general smearing or 'plastic' appearance of the picture; and/or horizontal jitter of the reproduced picture, especially at the top, where 'flag-waving' or incorrect hues may be displayed. To minimize these problems ensure that all VCRs involved are in good condition and correctly set up; that any copy/edit switch which may be provided is used; that the number of copy-generations is kept as low as possible – 2nd generation is acceptable on VHS, 3rd is acceptable on ED-Beta and S-VHS; and that all signal transfer takes place at baseband via AV sockets.

16

VCR CHROMA STAGES

Unlike most other chroma-processing stages in colour TV equipment, in a VCR the chrominance signal is not demodulated, decoded or encoded. The colour burst stays with the signal throughout the recording and playback processes, ensuring that their phase and frequency relationship is not disturbed. This means that such faults as incorrect hue, individual colours missing or distorted, and saturation errors are rare. In theory the first two cannot happen, and the third can only arise from an incorrect burst level relative to the rest of the chroma signal: a rare occurrence. This 'sealed signal package' system means that the chroma signal from a VCR tends to be either completely right; or missing altogether due to the action of a colour-killer either within the VCR or in the TV/monitor used to display the picture. Such exceptions as exist fall into the categories of coloured beat-patterns of various sorts, or a 'confetti' effect of chroma noise, with or without locked colour in the picture – these will be dealt with later.

The first essential in all VCR chroma fault-finding work is to ensure that any user switches (i.e. colour/mono, PAL/SECAM/NTSC colour1/colour2, etc.) are correctly set, and that the VCR is operating perfectly in black and white: this eliminates problems due to head defects, signal routing, pulses missing, servo errors, etc. This old adage from the early days of colour television is every bit as true of VCRs! The only circumstance where a 'common' component can give rise to chroma faults only is where the video heads have been transposed during replacement, i.e. the disc or drum fitted 180° out of true. Then the machine may well replay its own recordings in colour, but give black-and-white reproduction from other tapes, and make recordings which play back without colour in another machine. The effect on luminance depends on machine design.

In all cases of chrominance problems, first check whether the symptom arises on record, playback or both. This is even more important than for luminance troubleshooting, since the chroma section shares many functions between record and playback processes. If playback of a good tape is in mono-chrome, make a recording from a colour programme and try replaying it in a known-good machine; if the resulting picture comes up in colour a large area of the suspect machine's chroma system is exonerated. Where the fault (as is usually the case) is present in both record and playback modes, it is easiest diagnosed during playback of a pre-recorded colour-bar tape.

Some users complain of poor definition in the chroma signals from a VCR. This mainly shows on small captions and graphics (like those from a computer or the 'Text-in-Vision' transmissions during daytime) whose verticals may be smudged or colourless, while horizontals are more acceptable. This arises from the small bandwidth (500kHz) of the VCR's chroma circuits, and cannot be overcome.

No-colour faults

This section is addressed mainly to situations where the problem is present in both record and playback, indicating that the culprit lies in one of the 'common' circuit sections. For faults which are only present during record, or only present during playback, see later.

Figure 16.1 shows, in block diagram form, a common configuration of colour-under processing stages for VHS format. During record the chroma component is recovered from the composite video signal by filter A then bandwidth-limited to about 500kHz. This signal at 4.43MHz is now beat against a local frequency of 5.06MHz (point c) to produce a colour-under frequency of 627kHz for recording on tape. The 5.06MHz signal contains characteristics which are transferred to the recorded chroma signal: it is locked to incoming line frequency, and contains a 90° phase advance per TV line for head B signals only, identified by the SW25 (flip-flop) pulse train from the head-drum servo section. If these frequency and phase

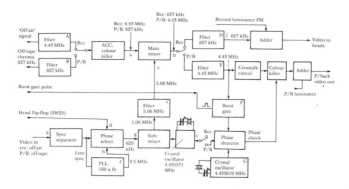

Figure 16.1 *Block diagram of the main processes involved in colour-under processing in VHS videorecorders*

characteristics are not correct during record there will be no colour on replay. A colour-killer circuit comes into action to switch off the chroma channel. Recording bias for the chroma signal comes from the f.m. luminance signal, whose amplitude is thus critical for good colour reproduction.

During replay the 627kHz off-tape chroma signal is intercepted by a bandpass filter B, then applied to the same mixer as was used for down-conversion during record. Again the local signal is at 5.06MHz, with a phase-stepping characteristic for head B's signals which restores normality to the 4.43MHz (selected by a filter E) output from the up-convertor. Next comes a two-line delay/matrix circuit for cross-talk cancellation and a colour-killer section which permits colour to pass so long as the phase of the outgoing chroma is in agreement with that of the local crystal oscillator G, to which it should be slaved by a phase-locked loop.

For tracing no-colour faults use an oscilloscope to check the off-tape signal – for colour bars it should resemble Figure 16.2. This should be traceable at the output of filter B in Figure 16.1, and at the input (point a) to the main mixer. Check next at the mixer output, point B. If there is no 4.43MHz signal here, it's likely that the local 5.06MHz input is missing from the mixer's second input, point c. Check for signal continuity through 5.06MHz filter C, then the conditions at the sub-mixer, both of whose inputs (625kHz and 4.435MHz) must be present to produce a 5.06MHz signal capable of penetrating filter C. If the 4.435MHz signal is missing, check its crystal oscillator; if the 625kHz input is absent the phase-switching chip is suspect so long as it is has operating power, and is receiving a SW25 feed; a

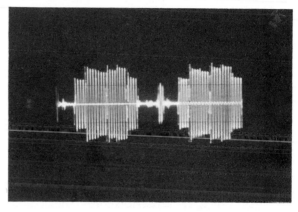

Figure 16.2 *Oscillogram of colour-bars at 627kHz. This waveform would be present at point j in Figure 16.1 during record; and at point a during playback*

sample of off-tape sync; and a 2.5MHz signal from the 160×fh phase-lock loop. Failure of the 2.5 MHz signal should lead to a check of the VCO and its phase-controlling input.

It may happen that a credible-looking 4.43MHz output is present from the main convertor, but that it is disordered in a way that is not immediately visible on the scope trace; this signal will not survive the colour-killer in the VCR or in the TV/ monitor. The points to check here are the *frequencies* of 4.435MHz and 4.4336MHz oscillators, and for the presence of flip-flop and sync input signals to the phase-rotation and PLL sections. To set up the crystal oscillators (they need to be within 20Hz of correct frequency to assure correct colour operation) an accurate frequency counter is required, hooked to a point where its input capacitance cannot upset crystal operation. A suitable test-point and adjustment procedure is given in the service manual. *Intermittent* colour is often the result of a crystal wandering in frequency or spasmodically shutting down.

If the chroma signal becomes weak, colour saturation is not affected until the level falls so low that the killer (on record or playback) is triggered to delete colour altogether. Up to that point the colours become progressively more grainy and confetti-strewn, particularly noticeable as a 'cotton-wool' effect on saturated blues. It is unusual to experience this on both record and playback, suggesting that one of the 'local' feeds: 5.06MHz, (check filter C), 4.435MHz, etc. is low in amplitude, or that one of the chroma bandpass filters – where common to record and playback – is partially O/C.

Unlocked colour

Sometimes the picture shows horizontal bars of incorrect hue, possibly on an intermittent basis as the colour killer comes in and out of operation. This is generally due to one of the crystal oscillators or the APC/AFC loop running off correct frequency – set up as per the manual and replace any crystal whose performance or stability is in doubt. Whenever a frequency counter is used to measure a waveform which has more than one component (i.e. the 5.06MHz phase-rotational signal) the reading cannot be relied upon. Such measurements are seldom necessary, but can be made by artificially holding the signal in one mode: to read 5.06MHz (point c in Figure 16.1) or 627kHz (point b) hold the head switching line, point d, either high or low, which can often be achieved by temporarily disconnecting the SW25 input. It is better in practice to check the source frequencies to the sub-mixer, which can, after all, only produce the beat product of its two inputs.

One-chip chroma processor

Figure 16.3 shows the block diagram of a single-IC chrominance processor as used in more modern practice. Record/replay switching is carried out electronically within the IC, governed by 'REC5V' and 'REC5V' control lines. The bandpass filters for 4.43MHz and 5.06MHz, and the 1MHz low-pass filter in the record signal path are all contained in the same assembly FL8001. The reference base for all chroma processes is the 4.43MHz crystal XC8001, which also provides a clock reference for the servo circuits, as shown at the bottom LH corner of the diagram. This circuit has no adjustments whatsoever, so that servicing action consists of checking for the presence of operating voltage VCC; correct-frequency oscillation of XC8001 at TP8001; the colour-killer input (low at pin 21); the key inputs SW25 at pin 5 and HSS (sync) at pin 54; and the chroma signal inputs at pins 32 and 34 for record and play respectively. If these are in order and the output at pin 17 – 627kHz for record and 4.43MHz for replay – is missing or disordered the IC is suspect. Before condemning or replacing it, compare the individual pin voltages with those given in the manual's circuit diagram to identify any faults in passive components external to the IC, typically leaky or O/C electrolytic capacitors, or physical faults in printed board conductors/connectors, plugs or sockets.

Betamax chroma circuits

In essence, the Beta colour-under system is similar to that for VHS. During record the phases of subcarrier signal for both A *and* B heads are rotated in opposite directions. Figure 16.4 shows that fault-finding processes and key test points are almost identical with those for VHS systems. During record the 4.43MHz chroma signal has a pilot burst added before being beat against a locally-generated signal of alternately 5.119MHz (head A) and 5.123MHz (head B). These alternating frequencies come from a sub-mixer whose inputs are (1) a 4.43MHz subcarrier locked to incoming chroma; and (2) an alternating 685/689kHz signal derived from a PLL working on incoming line sync. Frequency switching is triggered by the SW25 squarewave from the head servo section. Again, to read any of the 'alternating' frequencies it is necessary to 'stick' the SW25 input high or low as necessary.

During replay the constant-phase pilot burst (inserted into the chroma waveform at a point coinciding with the line sync pulse as shown in Figure 16.5) is used to lock the playback chroma signal to a local 4.43MHz crystal reference G. This oscillator and the alternate 685/689kHz PLL are used on both record and playback, so problems in either (i.e. crystal off frequency, loop broken) deletes colour on both. To check the operation of a VCO in a phase-locked-loop, use a low-impedance

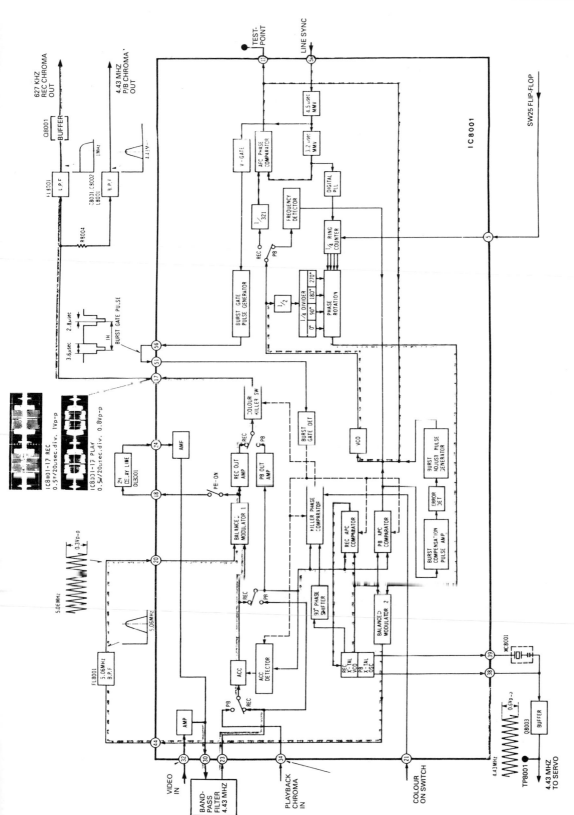

Figure 16.3 *One-chip colour signal processor (Panasonic)*

159

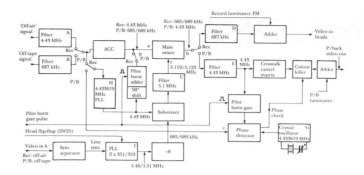

Figure 16.4 *The essential processes in the Betamax colour-under system*

external source of d.c. voltage applied to the error control line, e.g. points e and f in Figure 16.4. Start with a voltage corresponding to that of normal circuit operation. As the applied voltage is varied, a counter should show corresponding changes in oscillator output frequency. Similarly, as the VCO frequency control is adjusted (within the limits of the loop's correcting range) no variation in output frequency takes place, but the error voltage varies considerably in a correctly-operating PLL. These tests are useful in solving 'chicken or egg' problems in PLL loops generally.

So long as the key signal components 4.43MHz subcarrier, SW25 pulses, pilot burst gate pulse, and line sync are present, and the PLLs are locked, most faults will be confined to either record or playback (see later).

Video-8 chroma processing

Video-8 machines use no pilot burst, and have virtually the same principle of operation as VHS machines, so that the diagrams of Figures 16.1 and 16.3 are largely relevant, and the same key signal points and waveforms apply. The main differences between VHS and V-8 are:

1 The colour-under frequency, which for V-8 is 732kHz.
2 The phase-advance characteristic is applied to head A, while B's phase remains constant.
3 The burst signal amplitude is doubled by a burst-emphasis stage during recording and halved again during replay to combat noise.
4 The outer sidebands of the chroma signal are given extra emphasis during record, compensated

in the playback process. The frequencies for Video-8 (referring to Figure 16.1) are 732kHz at point a (playback), 5.16MHz at point c, 732kHz at point g, and 5.86MHz at point h.

In tracing no-colour faults, check that the crystal oscillator is running at the correct frequency; that SW25 and line sync pulses are present; and that 5.86MHz carrier is emerging from the VCO/PLL section.

Loss of colour on record

In a machine which proves capable of replaying a good tape in colour, but whose own recordings play back in monochrome, the problem is almost certainly confined to the upper blocks in Figure 16.1 or 16.4. Using a locally generated colour-bar signal, check with an oscilloscope that the 4.43MHz signal (oscillogram in Figure 16.5) comes correctly through filter A and the REC/PB electronic switch. So long as there are no problems in the a.c.c. or killer section, a clean and stable signal should be present at the main mixer input point a. Good replay will have proved that local signal is being applied to second input c, and that the main mixer is working. Trace the 627kHz down-converted chroma signal, Figure 16.2, through the electronic record switch, through filter D and thus to the 'add' block, where it is combined with the luminance f.m. record signal to render a waveform similar to those in Figure 16.6 for application to the heads.

Where a chroma signal can be traced all the way onto the tape, but replay remains in monochrome, the likelihood is that the 625kHz (etc.) PLL (Figure 16.1) or 4.433MHz (Figure 16.4) PLL is

Figure 16.5 *Colour-bar waveform at 4.43MHz – compare with Figures 16.2 and 16.7*

not locked to incoming sync. So long as replay is correct the problem is most likely to be in the 'off-air' video feed to the sync separator, points (k) in the diagrams.

For Beta machines the above signal-route tracing should also be carried out – the lettered points and block diagram layout in Figure 16.4 are purposely similar to those of Figure 16.1. For Betamax, an important point to check is the pilot burst shown in Figure 16.7. Check that it is around 80% of peak chroma level and if necessary (using a double-beam scope) that it coincides in time with the period of the line sync pulse. If it's missing or mistimed, check presence, amplitude and timing of the pilot burst gating pulse. If the pilot burst *is* present in the record waveform, its phase may be wrong – check the 4.43MHz PLL (block H in Figure 16.4) and the 90° phase shift network on its output route to the pilot burst adder stage.

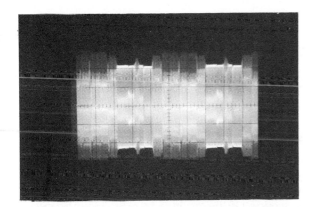

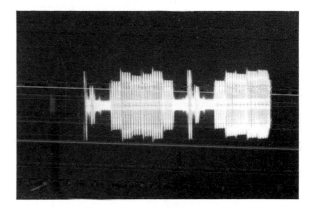

Figure 16.7 *Colour-bar waveform at 687kHz in a Beta machine. Note the large pilot burst signal just to the left of the 'real' burst*

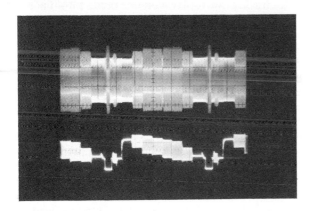

Figure 16.6 *The colour-under signal riding on the luminance f.m. carrier. (a) shows a VHS record waveform, and (b) a Beta one, with the CVBS waveform on the second trace of the oscilloscope*

For loss of colour during record in Video-8 machines, again follow the procedure given for the basic VHS system and the block diagram of Figure 16.1, bearing in mind, whenever checking with a frequency counter, the different system frequencies already given for V-8 format. The burst-emphasis stage in this system merits a special check: if the burst level going onto tape is wrong, no colour or *incorrect saturation* will be the result during playback because the a.c.c. system uses recorded burst level as a reference for regulating chroma amplitude. The burst emphasis gating pulse is derived from a form of delay circuit triggered by the off-air line sync pulse.

Regardless of the format, bear in mind that loss of colour on record *can* be due to a sub-standard chroma input signal where it comes from some local source such as a TV camera, computer, pattern/effects generator etc. Incorrect subcarrier frequency or burst level can be responsible, and the symptom is particularly misleading where the deficiency is such that the signal is tolerable to a TV or monitor (used for record-checking on the E-E signal) but falls outside the range of the PLL or colour-killer stage of the VCR. Check the integrity of the input chroma signal with an oscilloscope and if necessary a counter. This is particularly important with a camcorder, where it's difficult to apply an alternative input signal, and where in many cases the colour-under chrominance signal is produced *directly* from the camera electronics.

Other record-only colour faults

A common problem which appears to stem from the chroma recording section is a display of floating random colour bands on an otherwise good picture. It happens only when recording over a previously-used tape in a machine whose full-erase head is not working, generally due to failure of the erase/bias oscillator. This section is generally regarded as part of the *audio* circuitry, and for most machine designs the intruding colour bars are accompanied by sound problems – typically a strange and disembodied sound track from the tape's previous recording. More details in the next chapter.

Rarely, it can happen that during record only, the phase or frequency of the colour-under signal is incorrect, leading to unlocked bars of colour or 'cycling' incorrect hue on the played-back picture. For this, check the crystal(s) and PLLs, also any phase-shift networks concerned with burst or pilot burst in the chroma record section.

The colour-under section contains several filters as Figure 16.1, 16.3 and 16.4 show. If the filters fail by going open-circuit the result is a no-colour symptom on record, replay or both, fairly easily traced with an oscilloscope and block- or circuit diagrams. If, however, the filter fails internally or comes 'off-earth', its bandwidth may increase, leading to patterning in the off-tape picture; it is more likely to show on the *luminance* display of highly coloured areas of the picture than as a coloured pattern. Check by turning the colour control of the monitor right down and comparing the degree of patterning on the 'coloured' and grey

or white sections of a colour-bar display recorded by the suspect machine.

As already mentioned, the luminance f.m. carrier acts as recording bias for the chrominance signal on tape. Too low a luminance writing current shows best on the black-and-white picture, but excessive writing current can permit fully-saturated colours (i.e. large chrominance signal amplitudes) to run the tape into magnetic saturation. The effect is a 'crushing' of bright colours, best seen by comparing the amplitudes of burst and colour-bar signals in the replayed chroma signal. Too high a luminance writing current is most often the effect of maladjustment, but check (where relevant) the operation of any automatic colour/monochrome record-current switch, which boosts writing current when it thinks that no colour is present in the signal. Excessive *chroma* recording current can have the same saturation-limiting effect. Set it as instructed in the service manual.

No playback colour

Where a VCR records colour correctly (as proved by good replay in another machine) but cannot produce colour in playing back a known-good tape, the trouble is confined to those circuits which are used solely in replay. This may seem an obvious statement, but most of the function blocks in Figures 16.1 and 16.4 are thus eliminated: the two mixers, filter C and the bottom PLL in the VHS diagram, and both PLLs in the Beta diagram must be working if the machine records colour properly.

In tracing no replay colour, then, play back a colour bar tape and start with an oscilloscope at the input and output of 627kHz filter B in Figure 16.1. The waveform to expect is that in Figure 16.2, which should appear at main mixer input point a so long as filter B is intact, the playback switch is working and the colour signal is not so weak or corrupt that the colour killer has shut down. A 5.06MHz signal should be entering the main mixer on point c if recording is OK, and a somewhat 'messy' output should be present at point b – it contains many frequencies. Check it through the second REC/PB switch to the input of 4.43MHz filter E. A cleaner signal, looking like the waveform of Figure 16.5, will be present at filter E output if the filter has continuity. Trace it through the cross-talk-cancelling matrix to the replay colour-killer, where it may well disappear. Assuming for a moment that it *doesn't*, continue with the oscilloscope to the luminance/chrominance adder, and finally to the video output terminal and/or the r.f. modulator video input pin, looking for a waveform like that of Figure 16.8. If at any point in this chain the chroma waveform disappears the fault has been localized – an open-circuit filter, coupling capacitor or amplifier/buffer transistor is often responsible.

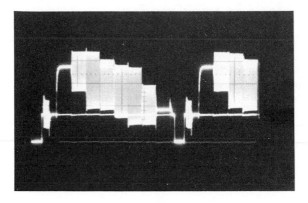

Figure 16.8   *Standard CVBS waveform for colour bars*

If the chrominance signal 'dies' at the replay colour killer, or if it appears to be leaving the VCR in r.f. or AV links but not surviving the TV/ monitor's colour killer, it's almost certain that the phase of the outgoing subcarrier is not locked to that of the local 4.43MHz crystal. In some cases (i.e. where the VCR has a colour/mono switch, or where the control line to the replay killer is accessible and its operating levels known) the killer can be overridden. If this action *does* restore normal colour, check the killer control line. More often, the result of overriding the killer is a rainbow of coloured lines on the monitor screen or still no colour due to the action of the monitor's killer, confirming the lack of phase lock in the VCR's up-conversion chroma circuit.

For this, again referring to Figure 16.1, check the operation of the burst gate F and if necessary the timing and amplitude of its gating pulse. Check that the reference oscillator G is running at the correct frequency, and that both 4.43MHz feeds are reaching the phase detector. If so, check its error output, and (if necessary with an external variable voltage source) that the 4.43MHz oscillator is responding to its control voltage input e. This completes the phase control loop.

For Betamax VCRs, the process of no-colour tracing is virtually identical. The blocks in Figure 16.4 are lettered to correspond. Here the pilot burst gate pulse coincides in time with the line sync pulse. Not shown in the diagram is the pilot burst cleaning stage, whose job it is to gate out the pilot burst once its usefulness has been realized.

In Video-8 chroma circuits, the same is true again, broadly based on Figure 16.1. With all formats, failure of the AFC loop (bottom left of Figures 16.1 and 16.4) will delete all colour from playback only under the unusual circumstance of a discontinuity of the path of the *off-tape* luminance signal to the sync separator – check for a video signal at points 'k' in the block diagrams.

Colour flicker

So long as luminance reproduction is correct, 25Hz (field rate) colour flicker is unusual. If the off-tape chroma signal amplitude at point 'a' in Figures 16.1 and 16.4 varies markedly between heads, poor l.f. response in *one* is indicated: check the pre-amps and their input tuning circuits. If the chroma levels from the two heads are reasonably similar, any 25Hz flicker problem is likely to stem from the phase-rotational circuit, the only other section of the chroma system which discriminates between heads; or the crosstalk-compensation circuit, see below.

Chroma crosstalk

The effect of crosstalk on the replayed chroma signal is similar to that of noise, but has a more regular and symmetrical 'pattern' effect, best seen and memorized by deliberately upsetting the delay-line matrix adjustments. If excessive crosstalk is suspected, the best approach is to follow the manufacturer's alignment instructions. If this fails to clear the problem the control's response (or lack of it!) gives useful clues to the trouble-spot. In such (rare) cases, dry-joints or a cracked delay line are typical culprits. In some VCR designs a faulty or maladjusted crosstalk-cancellation circuit can give rise to colour flicker; horizontal colour 'beats'; or colour line-pairing effects.

Coloured beat patterns

The chroma signal coming from the heads is very weak, and thus vulnerable to external interference on frequencies around 600–800kHz (350–500m

163

wavelength) which embraces the lower part of the MW radio broadcast band. The effect of break-through in areas near MF radio transmitters is usually blue/yellow or green/magenta diagonal lines which disappear when the monitor's colour control is turned fully down. If the colour patterning disappears when the aerial lead is removed from the VCR, try the effect of a *braid-breaker* or a high-pass filter in the aerial lead, as near as possible to the VCR. If the interference remains, the use of tinplate screens over the head-drum assembly and the head preamplifier section may provide a cure. Discuss the problem with the VCR manufacturer, who may have available a screening kit or modification instructions.

Other possible causes of coloured beat-pattern or herringbone interference on replay of a good tape are excessive luminance carrier leak due to faulty or maladjusted f.m. demodulator or chroma up-convertor circuits. Adjust any *carrier leak* and *balance* controls for minimum 'grass' on the video waveform during sync and blanking intervals. A more remote possibility for this symptom is a faulty bandpass filter in chrominance *or* luminance circuits.

Colour in trick modes

The performance of VCRs in such modes as still, cue and review varies tremendously between make, model, price and vintage. Before suspecting a fault in the chroma circuits, verify how good the colour *should* be in trick modes, then use the maker's block diagram and circuit description to check the half-line 'jump' and quasi-burst generator circuits fitted to some models.

17

VCR AUDIO SERVICING

The audio section of a VCR gives little trouble, and such faults as are encountered most often have their root in the head/tape interface, in the form of dirty or misaligned heads or 'mechanical' problems on the deck, many of which were covered in Chapter 12. With virtually any audio fault in a tape machine, the first essential is to thoroughly clean the head(s), whose position for a conventional (longitudinal-recording) machine is indicated as no. 7 in Figure 12.1, and in the tape-path diagrams of Figures 12.3 and 12.7. The audio section of the head assembly (Figure 12.13) is the upper part, embracing the very top edge of the tape ribbon.

It's important to understand the limitations of longitudinal sound recording systems to avoid fruitless investigations of the mechanical and electrical sections of a VCR whose audio performance does not come up the expectations of user or engineer. In general the sound channel of a conventional VCR is inferior to that of other domestic sound sources like f.m. radio, vinyl disc, audio tape and particularly compact disc. The frequency response is limited to about 10kHz (SP mode) and 5kHz (LP mode) which suppresses treble; the S/N ratio is at best 40–50dB, which means that hiss is noticeable in quiet passages; and wow/flutter levels can be high enough to be perceptible on sustained notes. All these problems are absent from the 'Hi-Fi' sound systems discussed later in this chapter.

In fault-finding, once again it is important to localize the problem to record, playback or common sections. For any faults which affect both record and playback, diagnosis is easiest done during replay of a 'work tape' recorded throughout with a constant tone of 400Hz or 1kHz. The only exception, perhaps, is the azimuth setting of the audio head, where the test tone from an alignment tape provides the best signal for adjustment.

Audio channel set-up

The sequence of adjustments for a longitudinal-sound VCR's audio section is given in Figure 17.1. Ensure first that the head is clean, and (if it has been changed or tampered with) that its height, tilt and azimuth settings are correct. Procedures for these checks were given in Chapter 12. Fault tracing now begins with test-tape replay, an oscilloscope, and a block diagram. In quoting set-up levels, manufacturers very often quote signal amplitudes in r.m.s. figures for use with an a.c. millivoltmeter. If this instrument is not available, multiply the figure given by 2.828 to arrive at a peak-to-peak figure for use with an oscilloscope – thus 15mV r.m.s. becomes 42.4mV p-p. Similarly, figures (especially at input/output and exchange points) are often quoted in dB. The standard level of 0dB corresponds to 0.776V r.m.s. (about 2.2V p-p) across 600Ω. The impedance figure is not taken seriously (!) so base conversions on 0dB = 2.2V p-p for a sine wave. Thus $-$60dB (a typical microphone input level) becomes 2.2mV p-p; $-$20dB 220mV p-p; and $+$10dB about 7V p-p. A decibel converter based on VCR practice is given in Figure 17.2, and

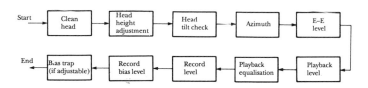

Figure 17.1 *Order of check and adjustment for complete set-up of the longitudinal audio section of a VHS or Betamax VCR*

V_{rms}	240 µV	780 µV	2.4 mV	7.8 mV	24 mV	78 mV	240 mV	780 mV	2.4 V	7.8 V	V_{rms}
V_{pk-pk}	700 µV	2.2 mV	7 mV	22 mV	70 mV	220 mV	700 mV	2.2 V	7 V	22 V	V_{pk-pk}
dB	−70	−60	−50	−40	−30	−20	−10	0	+10	+20	dB

Figure 17.2 *Ready-reckoner for audio levels quoted in VCR specifications*

a typical manufacturer's level diagram (these are very useful in fault-tracing) in Figure 17.3.

Some guidance on eliminating various areas of the machine's audio system is given in Figure 17.4. If another VCR is not readily available to check suspect recordings, the golden rule is to get replay correct first to eliminate common factors such as the audio head/tape interface, the r.f. modulator and the power supplies to the audio section of the machine.

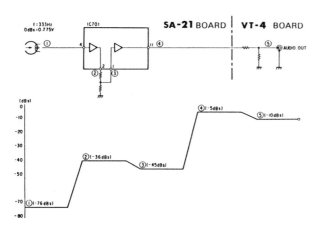

Figure 17.3 *Sound level diagram for a typical VCR (Sony)*

Replay faults

With a good tape inside the machine, the audio head newly cleaned, and the r.f. modulator proved OK by its ability to handle E-E sound in stop or record mode, the first thing is to decide whether the problem is electrical or 'mechanical' in nature. If there is no or weak sound, touch a finger to the audio head connections. A loud buzz now indicates that the head is faulty or, more likely, that it's getting no signal from the tape: check head height and head/tape contact. If the 'finger-test' produces no buzz, the electrical circuits are faulty (see below).

Many replay problems are caused by incorrect tape operation. The effect of wrong azimuth setting is a loss of treble response, except on the machine's own recordings. Weak playback audio (often intermittent and fluctuating) with high noise level is the result of insufficient tape pressure against the head – check tape back-tension and audio head tilt (zenith) as described in Chapter 12. If these and head-height setting are correct, the head may be worn and in need of replacement; this diagnosis can usually be confirmed by close examination of the head's front surface, which may appear 'flattened' and have a visible groove or step at the edge of the tape path. The groove can also be felt with a fingernail.

Flutter on sound is caused by short term variations in tape speed or tension. Check for eccentric running of the video head disc or drum, then if necessary for smooth running of the tape at the video head-exit guide and the next fixed guide downstream. If these are in order, the audio head itself may be responsible (surface friction), but more likely is a problem with the capstan or pinch roller. Rotate the capstan by hand to check for 'grumbling' bearings in its motor, and spin the pinch roller likewise – it should be free on its shaft. Sometimes replacement of the pinch roller cures flutter.

More often encountered than flutter is *wow*, a cyclic variation in the pitch of the replayed audio. To qualify for the 'wow' tag, the rate of variation must be below 20Hz. In VCRs it is usually much slower than that: 0.5–5 Hz is typical. Wow is caused by speed changes in the tape's progress, so the most obvious suspect is varying capstan speed. First, clean the capstan and pinch roller surfaces, then check that the pinch-roller pressure is approximately correct at about 2.5kg; and that the take-up torque is not excessive – see Chapter 12. If these are in order, useful clues can often be gleaned by analysing the wow-rate in terms of rotation-rate of capstan, pinch-roller, or take-up spool. Again, manual turning and 'feeling' of capstan and pinch roller (machine switched off) can reveal any tight spots or similar defects. Very often, a replacement pinch roller cures wow problems

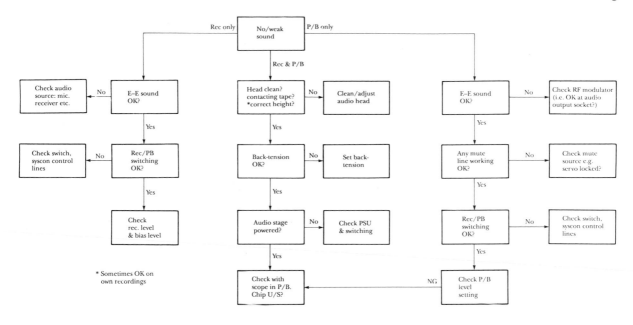

Figure 17.4 *Audio fault finding in 'longitudinal' sound systems*

even when the old one looked, felt and ran OK. Where no mechanical cause for it can be discerned, wow and/or flutter may stem from a servo or motor fault (details in Chapter 14). If a wow/flutter meter is available, a 'normal' level may be regarded as 0.25% to 0.5% w.r.m.s.

Even away from the deck, 'mechanical' problems in the form of poor switch contacts can upset sound reproduction. Intermittent sound, hum or crackling arises from dirty rec/play slider switches in old machines; and particularly from defective head-switching or signal-routing relays where fitted in later machines. Apart from the symptoms listed above, 'motor-boating' and instability/oscillation can be caused by a few ohms of contact resistance in a relay. Cleaning rarely provides a lasting cure, so replacement is recommended.

Purely electrical faults in the sound replay circuits are quite rare. Figure 17.5 shows a block diagram of a typical 1-chip audio system for a VCR. For complete loss of replay sound check that the IC is powered via Q704, and that the mute lines into D8 and IC pin 10 are not high. The signal entering the chip at pin 4 is too small to easily detect (see Figure 17.2 and Figure 17.3) but should be seen to emerge from IC701 pin 2 at about 34mV p-p, and at IC pin 11 at say 1.3V p-p. If these levels are incorrect, check the a.g.c. components fitted to IC

pin 15, then suspect the chip itself. Most circuits have replay audio level control presets, and some have 'discrete' audio amplifiers. Replay equalization is always shaped by negative feedback loops containing R and C elements, and are seldom adjustable. Incorrect equalization causes tonal imbalance in the sound, but remember the constraints described at the beginning of this chapter!

Very weak replay sound, with high hiss level, suggests a faulty head or a problem in the early stages of the replay amplifier – in discrete designs, check the resistors and transistors at the amplifier input. *Distorted* replay sound is rare: use an oscilloscope to identify the stage at which a sinc-wave signal from a test-tape becomes corrupt. Possibilities are failure of a negative feedback loop, incorrect biasing, or semiconductor failure.

Record faults

Audio recording faults which also have an effect on playback have been largely covered in the preceding paragraphs. Those symptoms which stem from head/tape signal transfer problems or deck faults are *doubly* bad when the machine replays its own recordings – giving useful clues to the cause of the trouble. The exceptions to this rule are head-

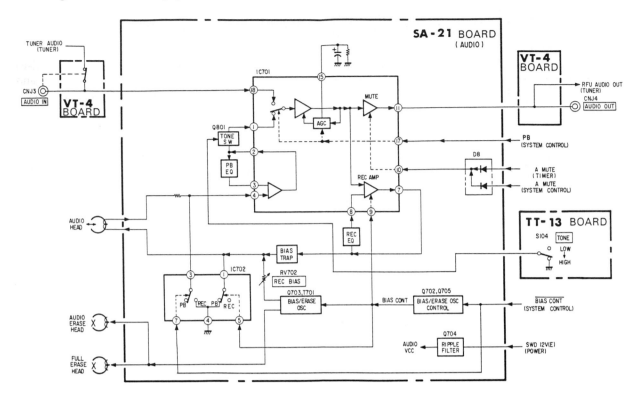

Figure 17.5 *Audio signal-processing stages (Sony)*

azimuth errors, constant tape speed errors and tape or head *height* faults, where the errors are concealed during replay of the machine's own recordings. This section, then, concentrates mainly on faults which are present only on record, and show up during replay in their own or another VCR.

The main factor in audio record is the *bias* signal, which is generated at typically 60kHz by a power-oscillator, then added to the audio signal on its way to the recording heads. It keeps the signal variations within the linear part of the tape transfer characteristic, and its level is thus critical. As the bias is wound up beyond the optimum point the audio signal recorded on tape becomes deficient in treble, then weak, noisy and 'clipped'. At excessive bias levels, virtually no sound is recorded. A deficiency in bias level, on the other hand, permits distortion of the recorded signal, showing as 'kinking' at the zero-crossover points of a recorded sine-wave. If the bias signal is completely absent, no recording is made. Because the oscillator almost invariably provides both recording bias and erase power to the FE and AE heads, see Figure 17.5, useful clues are

available from the faulty recording. If it shows blobs of floating colour, and an old sound track (unconnected with the picture!) is heard, the bias/erase oscillator itself has stopped. In a set-up like that of Figure 17.5, check the BIAS CONT input, Q702 and Q705. If, however, the old recordings are being erased when the fault is present, check the setting and condition of the REC BIAS control and associated components. Typical p-p voltage level at the FE head is 40–60V p-p.

Setting of the recording bias level (here RV702) depends on the type and condition of the audio head, as well as the type of tape in use. For this reason, manufacturers often specify a 'trial and error' setting up procedure, in which *replay* levels of typically 333Hz and 7kHz recordings are compared; this time-consuming process is best, however. Alternatively, production audio heads may be individually coded, with different bias levels specified for each. The bias current is measured as an r.m.s. voltage across a resistor permanently fitted in series with the head. Multiply r.m.s. values by 2.828 to arrive at a p-p figure for scope use, and expect to need a sensitive scope and 1:1 ('straight-through') probe, since a typical reading may be 5mV. In all cases, follow the manufacturer's instructions to the *letter* – the bias level is very critical. Figure 17.6 shows a voltage waveform at the audio head during record of a test-tone. The actual signal is quite small compared with the bias waveform.

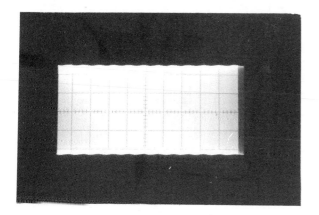

Figure 17.6   *Head-drive waveform during record of a sine-wave at 1kHz*

No- or weak-record faults, where bias and audio input signals are present and correct, is merely a matter of signal-tracing throughout the record amplifier, with the help of any block (Figure 17.5) and level (Figure 17.3) diagrams in the service manual. Once the faulty stage is identified, component checks can follow.

Check particularly the setting of the record-level preset (where fitted – the level is fixed in the circuit of Figure 17.5): too little drive to the head gives weak and noisy recordings, while excessive drive causes distortion, seen on a scope during replay as a clipping or flattening of the top and bottom of a recorded sinewave.

With camcorders and cameras a common user-complaint concerns changes in background noise level, depending on the ambient sound level during recording. This is a characteristic of the ALC (automatic level control) necessarily used to squeeze a natural dynamic range of say 90dB into the span of (e.g.) 40dB which the VCR's sound system can handle. Modification of the ALC characteristic is impractical, and the effect has to be accepted as inherent in the system.

Hi-Fi sound systems

An alternative audio record/playback system, Hi-Fi sound, is fitted to the more expensive VHS and Beta VCRs, and to all Video-8 format machines. Like the vision signal, here the audio signal is frequency-modulated onto a carrier wave which is then recorded with high writing speed into helical tracks in the tape. They follow the same paths as the video heads themselves. This confers very good sound performance, and is reliable, but when things *do* go wrong there is much more for the service engineer to know and check.

It is not our purpose here to explain the principles of operation of the system in any but the briefest terms, and then purely from the point of view of fault diagnosis. Comprehensive descriptions of the principles involved are available in manufacturer's literature, and in *Television and Video Engineer's Pocket Book* by the present author.

Figure 17.7 shows the bones of a Hi-Fi recording system. It represents one channel (L or R) of a stereo system (VHS, Beta Hi-Fi) or the single (mono) channel of a Video-8 machine. For record the incoming audio signal passes through a filter which removes frequencies above 20kHz (VHS, Beta) or 15kHz (Video-8). In the op-amp the signal is logarithmically compressed in dynamic range so that an input range of 80dB appears at S3 with a range of only 40dB, the result of precision control of negative feedback by the *compander* stages inside the dotted box. The audio signal now passes through a level control (set deviation) for application to a VCO, which acts as frequency modulation stage, in like fashion to the modulator in the vision record circuit. The differences here are that the carrier frequency is lower, and the deviation range narrower.

For VHS, frequencies are 1.4MHz for LH channel, 1.8MHz for right, with deviation of ± 150kHz; for Beta 1.44 and 2.10MHz respectively, deviated to ± 200kHz; and for the Video-8 format's mono sound channel the carrier is at 1.5MHz with deviation of ± 100kHz. In stereo (VHS, Beta) systems the audio carrier for the other channel is added to the sound f.m. signal on its way to the audio head-pair on the drum. In Video-8 machines the sound f.m. carrier signal is added to the vision (Y-f.m. + C) signals and recorded by the *vision* heads as an integral component of the 'vision spectrum'.

On VHS and Beta replay the tiny signals from the audio heads are separately selected by a

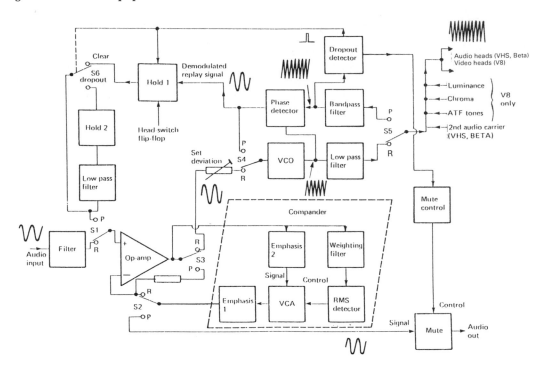

Figure 17.7 *Outline of Hi-Fi signal processing, representing one of the two channels available in VHS or Beta formats*

(delayed) SW25 switching squarewave and then picked out by filters tuned to the two carrier frequencies. In Video-8 the sound carrier is recovered from the vision signals by a sharply-tuned 1.5MHz filter. The VCO previously used as recording f.m. modulator now becomes part of a PLL demodulator, whose output represents the baseband audio signal, still in range-compressed form. For detection and concealment of dropouts and head-switching transients two hold circuits are now encountered, after which the 'patched' signal re-enters the op-amp via record/play switch S1.

On replay the op-amp operates at fixed gain due to the *fixed* resistor between its inverting-input and output terminals, so the signal passes unmodified into the compander block, which 'stretches' the signal back to normal, amplitude wise. It operates in exactly the same way as on record, but is now in the direct signal path rather than a negative feedback loop. The audio signal present at the 'P' pole of S2, then, is an exact replica of that recorded, and passes out through a mute stage (operated by the dropout detector) to become the replay audio output of the VCR. It is also applied to the r.f. modulator after addition to the other (LH, RH) channel to form a mono signal. VHS and Beta

VCRs always have an audio record and replay 'back-up' system in the form of the longitudinal audio track (which may be mono or stereo), and in the event of bad corruption and dropout, the replay signal line is automatically switched to this to maintain sound, albeit in 'Lo-fi' form. The switch-over is signalled by front-panel LED indicators on the VCR.

Fault diagnosis and setting-up

The first essential in servicing Hi-Fi audio systems is to clean all the heads on the drum, and ensure that the *video* tracking is spot-on, with a good vision envelope pattern on replay of an alignment- or work-tape. Tracking for Hi-Fi sound is more critical than for vision because of the narrower tracks of the former; this has led to the provision in many machines of an audio tracking indicator to enable the user to adjust the tracking control to optimum. The effect of incorrect tracking is intermittent drop-out and break-up of the sound on both channels. The audio f.m. envelope pattern can be picked up at a point before the limiter in the head amplifier circuit, often labelled 'AFM out' (replay).

With the tracking correct, get the sound right in replay before checking record functions – this eliminates the many 'common' circuit blocks in Figure 17.7. The waveforms to be expected are indicated on the diagram; amplitudes vary between models, but are given in the service manuals. A great advantage of servicing this type of circuit in VHS and Beta machines is the presence of two identical and independent channels, L and R, to give a ready-made, on-the-spot reference to correct waveforms, voltages etc., during fault-tracing. If *both* channels fail in the same way, the problem will be in one of the few components/feeds common to both: rotary heads, recording/playback head amplifier, muting, rec/pb switching, IC power supplies, etc. The one exception is where the audio heads are beginning to wear, when the RH channel (highest carrier frequency) suffers from excessive drop-out, especially during replay of 'own' recordings.

Adjustment of the headswitch point is critical. If it is incorrectly set, the result is 'beat noise' in playback, when the sound is marred by a 50 or 25Hz buzz or 'rattle' effect; the same symptom can arise from failure of the dropout/hold circuit, or from replay envelope faults. Bear in mind that the audio heads *both* work, for *both* channels during record and playback: one head 'out' gives 20ms gaps in the audio P/B envelope pattern, see Figure 17.8. Set up the headswitch points critically as instructed in the manual.

Incorrect replay level gives imbalance between L and R channels, but make sure that it is *true* imbalance, and not incorrect setting of the level indicators on the front panel. For other cases of one channel absent, weak or distorted, work through the block diagram with an oscilloscope, comparing levels and voltages with those published by the manufacturer, or at equivalent points in the good channel. When the offending *stage* is found, use the circuit diagram, scope and voltmeter to identify the faulty component. Sometimes it may be necessary to identify and override the replay mute line in order to trouble-shoot the Hi-Fi replay circuits.

Once correct operation of Hi-Fi replay from a known-good tape (some early mass-duplicated tapes from hire libraries cannot be relied upon, Hi-Fi track wise) is established, much of the electrical circuit, and the deck mechanics, guides, heads etc. are proved OK, and any record faults can now be tackled.

Again, the presence of two identical channels is a great help in setting-up and fault-finding. Apply a monaural input signal so that both channels have identical inputs, and use a scope to trace the signal, in baseband or f.m.-modulated form, from input terminal to recording heads. There are more adjustments in the record section compared to playback, and for complete setting-up you need an oscilloscope, frequency counter and possibly an a.c. millivoltmeter, though conversion of r.m.s. to p-p values can be done by multiplying by 2.828.

The symptoms arising from various faults depend considerably on machine type, state of wear of heads, tape type etc., but some possible fault patterns, with suggested causes, are given next:

1 Record current incorrect: both channels affected, but R may be worse due to its higher carrier frequency. Effects are very critical tracking adjustment, excessive dropout of sound, 'rasping' intrusions, intermittent muting.
2 Carrier frequency wrong: one channel only carries distortion, especially on loud passages. The other channel may be marred by 'birdy' sounds.
3 Deviation incorrect: imbalance between L and R channels, incorrect replay audio level in Video-8 machines. Excessive deviation causes distortion (clipping) of loud passages in the afflicted channel, and perhaps picture patterning in luma and/or chroma images in Video-8, dependent on sound level. In VHS, excessive deviation of one audio carrier can additionally cause 'birdies', distortion or break-up of the *other* channel due to crosstalk in the sideband spreads of the closely spaced carriers.

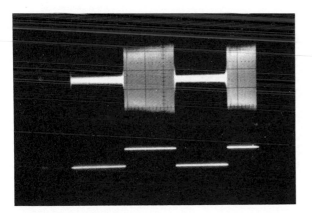

Figure 17.8 *FM sound carrier with one head defective. The squarewave below shows the* audio *head switching line*

Hi-Fi sound processing circuit diagrams inevitably look complex, and can be somewhat daunting, especially as they represent an area not as well known as video and other sections. Identify and check the key points and waveforms, and bear firmly in mind the effect of the various mute and switching lines, working initially from block diagrams.

PCM sound

The ultimate sound system is represented by Pulse-Code Modulation, as incorporated in the most expensive Video-8 camcorders and homebase VCRs. *All* Video-8 machines are fitted with the mono AFM recording/playback system in which an f.m. sound carrier at 1.5MHz is added to the vision signal during record, then recovered and demodulated at replay. PCM-equipped machines have an additional high-quality stereo sound processing stage in which each audio channel, L and R, is separately processed and digitized for recording on tape. The audio data is stored in a memory, typically consisting of two 16K RAMs, for fast readout in a 3ms 'burst' once per 20ms. This data-burst is gated to each *video* head in turn, while its fellow is writing the bottom part of the picture (Y + C) onto tape, so that a 'forward-extension' of the video tracks is used for PCM data as shown in Figure 17.9; this involves an extra 30° of tape wrap which, with the required 'head-overlap' period gives a total tape wrap angle around the head of about 220° as shown on the left of the diagram.

The electronic processing involved in PCM recording is outlined in very simple form in Figure 17.10, where it can be seen that virtually all the stages used in record are re-used during playback – another case for getting it right first on replay of a good tape! The incoming signals, L and R, are separately compressed in compander stages like that in the dotted box of Figure 17.7 to reduce their dynamic range, then *alternately* digitized in a D-A convertor whose sampling clock runs at 2fh, 31.25kHz. Emerging from the D-A convertor is a real-time serial data stream in 10-bit form, whose *presence* is important for servicing checks, but whose characteristics cannot be analysed by scope examination.

After conversion to 8-bit data (check that data is present) the binary signal enters an LSI chip associated with one or two memory ICs. Its operation is complex, and the main actions in tracing problems consist of checking that it has correct inputs and outputs: operating voltage; serial data in; system clock pulses; mode control input from the system control section, typically on four parallel data lines; parallel data transfer to and from the RAM memory(s) on typically 11 address and 8 data lines; a 2.9ms PCM gating pulse derived from tacho pulses via the head drum servo; and data out to the record f.m. modulator (where it is not incorporated in the same chip). The digital data is modulated onto an FSK (frequency shift keying) carrier for commitment to tape: 2.9MHz for digit zero, 5.8MHz for digit one. This carrier can be examined as a 'burst envelope' in the video-head amplifier channel during record and playback.

In playback the FSK envelope pattern coming

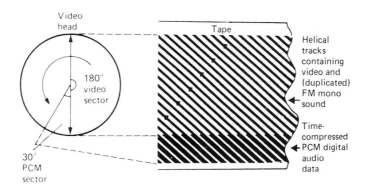

Figure 17.9 *PCM sound recording: the video heads read or write digital audio data for about 30° of drum rotation before being switched to the video channel*

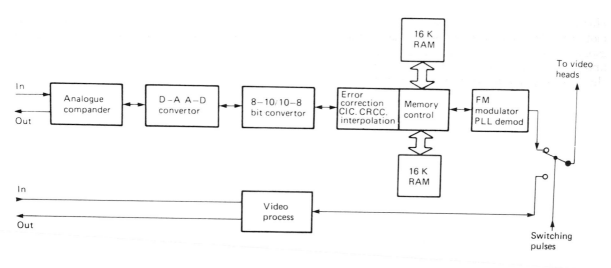

Figure 17.10 *PCM audio signal processing and video head switching*

off tape is gated out from the vision signal by normal head-switching; and by a 2.9ms 'enable' pulse. It is demodulated in a PLL incorporating the record-modulator VCO, then fed into the same memory chips as were used during record. In the LSI memory-control chip the data is reconstituted into a real-time 8-bit serial data stream, a key signal check-point in fault-tracing. A conversion back to 10-bit data follows, in which the two audio channels, L and R, are disentangled and presented separately to the companders which were used during record – like that in Figure 17.7, they operate in the direct signal path (rather than a negative feedback loop) to fulfil their *expander* role on replay. The expanded and reconstituted audio signals are now switched to the audio output sockets and r.f. modulator so long as they are not corrupt, as judged by a drop-out detector monitoring the replay FSK signal off-tape. Deterioration here invokes switchover to the AFM replay channel.

In fault-finding, replay an alignment or (preferably) well-recorded test tape, and use the scope to trace the signal from the output of the head pre-amplifiers forward. The digital signal is quite robust, but (as off-tape envelope inspection will show) can become hopelessly corrupt if the tape-entry guide is incorrectly set. The two channels are inextricably mixed together throughout the PCM processing, storage and tape sections, so any fault which affects only *one* channel will be in the small area of circuitry on the analogue side of the D-A/A-D convertor. Work closely with the manufacturer's block and circuit diagrams, checking key data and control transfer points as described above. Faults, in order of likelihood, may stem from tape/head interface and 'mechanical' problems; lack of control-microprocessor or memory-control chip reset or clock pulses; and 'stuck' data lines. Do not attempt any work on the PCM section unless the picture reproduction, in record and playback, is exactly right.

18

CAMERA SECTION

Home video cameras have developed rapidly over the years. Starting as separate units with multicore cord connection to a portable VCR, they have now become an integral part of a camcorder. Because of the complexity of their circuits and setting-up procedures, and their rarity compared with TV sets and VCRs, few service departments can afford to invest in the equipment and expertise necessary to *fully* service a wide range of makes and models of camera/camcorder. The cost of such equipment as light boxes and test-transparencies, vectorscopes and the many jigs and special tools required for each individual model makes 'casual' servicing not viable unless the faults are relatively simple ones. Once a service department *is* geared up with expertise and full service equipment, it needs a large throughput to justify the investment and running costs. Camera servicing, then, tends to devolve into two categories: 'first-line', in which the majority of faults encountered can be diagnosed and rectified; and 'specialist', involving replacement of pick-up tubes and subsequent setting up, alignment of colour encoders, etc.

This chapter addresses itself to first-line servicing, and regards mainly tube-type cameras. Solid-state image sensors (CCD, MOS) will be described later. They have fewer adjustments, especially after replacement of the sensor itself, but with all sensors you need to be sure of the diagnosis – replacements cost several hundred pounds! In all dealings with camera electronics and optics, closely follow the manufacturer's circuit and block diagrams in fault-tracing, and instructions during setting up and adjustment. Above all, *never* indiscriminately 'twiddle' pre-sets! Only adjust a pre-set when you know exactly what it should do, and have the means of setting it correctly. Even then, it's wise to precisely mark its original position so that the factory setting can be exactly restored if necessary.

Many sections of the camera are inter-dependent, and an idea of the order of functions, and their adjustment, is given in Figure 18.1. As with several other topics dealt with in this book, it is essential to

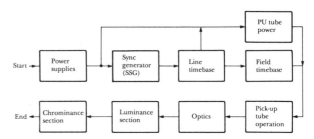

Figure 18.1 *'Dependence tree' for the stages in a colour TV camera or camcorder*

'get it right in black and white' before getting involved in the colour circuitry.

Camera block diagram

An outline of the individual stages and processes in a tube-type camera is given in Figure 18.2. For a completely inoperative camera, first check that it is powered by pressing the zoom button (lens barrel moving?) or looking for indication lights and viewfinder raster. If these are absent, the internal PSU and fuses should be checked. Typically a simple series regulator or mini-chopper circuit (see Chapter 3) provides two stabilized lines, e.g. + 9V and + 5V. Check, where necessary, the fuses and circuit protectors fitted: service manuals often give a helpful power distribution diagram, in which the regulators, fuses, protectors etc. are shown in relation to the sections they serve. If a safety resistor or fuse has blown, check the appropriate section: jammed lens motors or their drive transistor/ICs S/C, incorrect polarity of applied power and S/C line output transistors are the most common causes.

SSG generator

The heart of the camera's electronics is an SSG (Sync and Subcarrier Generator) which, based on a

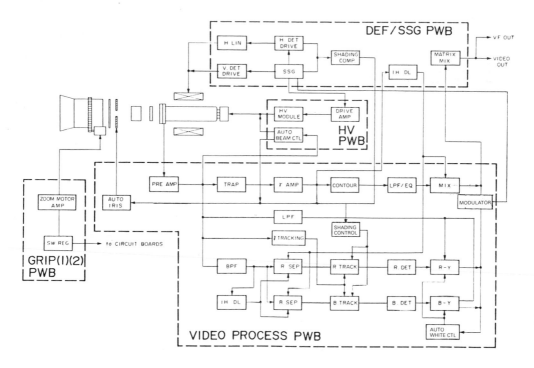

Figure 18.2 *Simplified block diagram of a video camera (JVC)*

reference crystal of typically 17.73MHz (4×fsc), produces correctly timed pulse trains for all parts of the camera: timebase trigger, blanking, syncs, colour encoding etc. If the SSG is not working, there will be no output at all from the camera's video socket, though sound and zoom motor will probably work. If the SSG is dead, check its operating voltage supply, its crystal clock and for excessive loading on its output lines before suspecting the chip; if the picture is absent but sync pulses are present at the video output socket, the SSG is probably functioning correctly. Correct clock frequency is crucial: adjust the master crystal's trimmer for 4.433619MHz subcarrier frequency, bearing in mind that an error of more than 20Hz here may lose you the colour altogether. Use a counter which has been checked for 4.433619MHz on the subcarrier generator of a TV locked to a broadcast transmission.

Pick-up tube

There are several types and sizes of pick-up tube in use: Saticon and Newvicon are the most common,

differing from the basic Vidicon type in their faceplate materials. In general, they have a cathode, a 'grid' and five 'anodes'. In a typical camera the voltages to be expected are as follows. Cathode 20V/blanking pulse; grid approx − 30V; G2 300V; G4 (variable with focus control) 150V; G5 approx 500V; and G6 (mesh) 1000–1400V. Another essential feed is heater voltage, which varies between 3V and 6.3V with tube make and type. All these voltages must be present and correct to produce pictures at all, and once the PSU and SSG are proved working the tube operating voltages should next be checked in cases of no picture. Most of them are derived from rectifier/capacitor sets fed from the flyback or line output transformer, so a short-cut can be made (Figure 18.1) by first checking whether the line timebase is operating.

Line timebase

The line scanning stage in a TV camera is a mini version of those used in TV sets and monitors, and covered in Chapter 4. The main differences are:

175

1 The line oscillator is replaced by, or phase-locked to, the SSG.
2 The energy requirement (for scanning and beam current) in a camera tube is very small.
3 The line scan must be very linear indeed in order to correctly recover the colour signal from the target's output.

Special linearizing circuits are used, and set up by reference to the *chroma* output signal: details later, and specifically in the maker's set-up instructions. In other respects the line scan stage is conventional, and subject to the advice given in Chapter 4. The chrominance signals from the target depend on high-frequency carriers, which can only be produced by a very *fine* scanning beam in the tube. The electrostatic focusing control (tube anode 4) is adjusted, then, for maximum *colour*. Since colour in this context means R-Y and B-Y, in practice the focus control is adjusted for minimum *green* on the screen of a monitor whose colour is turned fully up while the camera looks at a plain white sheet. A bright green picture is a common problem, and is generally due to incorrect camera beam focus. Green *highlights*, however, spring from a different cause; see below.

Field timebase

Again, the scanning energy required of the camera's field timebase is very small, but correct linearity and scanning amplitude is very important, not only from the point of view of obtaining a correctly proportioned image on the monitor screen, but also for correct rendering of colour. The source of flyback-trigger pulses is once again the SSG section, and many cameras have a scan-fail protection circuit in which the beam current in the pick-up tube is turned off in the event of field-scan-current failure. If the beam is not suppressed the symptom of this fault is a series of faint blurry vertical black and white lines. Switch off directly to avoid damage to the tube. In servicing and setting up camera scanning circuits, bear in mind that what you see on the monitor is a 'mirror-image' of the fault, so that, for example, loss of scan gives an 'infinitely' wide or high image; 'insufficient-width' scanning of the pickup tube face gives an excessively-wide picture on monitor; and 'cramping' of the bottom of the pick-up tube's vertical scan causes a stretching effect at the bottom of the picture as viewed. Until the geometry of the pick-up tube scanning is

exactly right there is poor colour; if scan characteristics are badly wrong no colour at all is produced.

Luminance chain

So long as the tube is energized, light is reaching its target and the beam and target voltages are correctly adjusted, there should be a signal coming off the tube's target connection ring at the front. The connection is a very short one to minimize capacitance, and leads to a high-gain, low noise preamplifier, generally using an FET at the input stage. One way of roughly checking the luminance chain (and thus eliminating the pick-up tube, iris, etc. from the 'no-picture' diagnosis) is to approach the preamplifier input point – or the exposed faceplate of the tube – with a finger if they are reasonably accessible. If no noise and patterning is visible, the luminance chain is broken somewhere. Trace downstream with a scope or signal injector.

From a servicing point of view, the luminance chain consists of amplifiers; a gamma-correction stage; a contour-enhancement section in which vertical (and in some cases, horizontal) edges are crispened; a clamp; an a.g.c. stage to compensate for varying light levels; and an add stage wherein the chroma signal and sync pulses are united with the luminance signal to render a standard CVBS output. The luminance chain is set up by reference to a test-chart containing a double step-wedge as shown in Figure 18.3(a), which when viewed on a

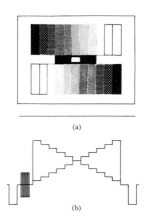

Figure 18.3 (a) *Double step-wedge test chart. The resulting video waveform is shown in* (b)

scope triggered at line rate gives a waveform like that of Figure 18.3(b). Gamma correction is carried out by varying the gain of the luminance amplifier according to the level of signal, thus pre-distorting the relationship between light level and electrical signal to compensate for the non-linear characteristic of the picture-tube used for display. The gamma (γ) characteristic required is 0.45, and the chroma amplifiers are also under the influence of gamma correctors. The effect of incorrect gamma is poor tonal rendering of the greys in the picture; that of incorrect a.g.c. control is low contrast or severe overloading, possibly with a negative picture; and of incorrect clamping (check for clamping pulse from SSG) a loss of stable black level, and brightness drift with picture content. These problems should be dealt with without regard to the chroma content of the picture, which will be incorrect while they are present.

Optical section

The lens assembly focuses the scene on the faceplate of the pick-up tube. For correct calibration of the distance graduations on the lens body, and for maintenance of good focus throughout the zoom range, the spacing between the rear of the lens assembly and the faceplate of the tube is crucial. This, the *back-focus* adjustment, is not difficult to carry out, and generally requires no special equipment. Get the iris wide open by fitting ND filters or by operating the manual (BLC) iris control. Focus on a distant object with the zoom lens in full telephoto position, then zoom back to wide-angle position and adjust the back-focus screw for correct focus. Repeat the process as many times as necessary until good focus is maintained throughout the zoom range, always adjusting the front focus ring in telephoto position and the back-focus screw in wide-angle position. Finally seal or lock the back-focus adjustment screw.

The iris itself forms part of an a.g.c. loop governed by feedback from the camera's video circuit on the amount of light entering the camera. Normally the iris is fully closed in the 'rest' (unenergized) position, a point to bear in mind when the symptom is 'no picture' and a degree of noise can be discerned on the monitor's raster. The state of the iris can be seen by looking into the lens – it should be seen to open when the camera is switched on. An oscillating iris (giving picture flutter during its long settling-down time after

scene change) should direct attention to its in-built *damping coil* and associated circuits.

Auto-focus systems vary tremendously with camera design, the basic concept being that the front of the lens barrel is driven round via a slipping clutch and reduction gearing from a tiny motor, all part of a servo loop whose positioning reference is the sharpness of detail at picture centre. Problems are rare, and may range from no-go at all to instability and oscillation of the lens. General information on servo loops was given in Chapter 14, but be guided in servicing by the manufacturer's block diagrams and circuit description.

Marks and blemishes on the picture will not be due to defects or particles on or in the lens assembly – anything here is way out of focus on the picture, and would be seen to rotate with optical focus adjustment. The tube faceplate *is* in sharp focus, so particles on its surface will show up clearly. The same applies to damage to the target layer, generally brought on by excessive (or spot/line concentrated) tube beam current; exposure to very bright lights, e.g. the sun; or by long exposure to a fixed scene containing bright spots. Mild target defects can sometimes be 'washed out' by leaving the camera running for many hours watching a bright plain white surface like a well-lit white card or the north sky. If the problem is not thus cured or sufficiently diminished, the tube must be replaced.

Worn or defective tubes are most often identified by an inability to set up beam current and target voltages (*dark current*) within the specifications set by the manufacturer in the service manual. Before suspecting the tube itself ensure that all electrodes, including the heater, have correct operating voltages.

No-picture camera fault-finding procedure is outlined in Figure 18.4.

Chroma processing

The chroma stages of a TV camera devolve into three distinct areas: *recovery* of the colour signals corresponding to R and B from the tube's target output signal; *correction* of the colour signals to compensate for shortcomings in the target surface and in the scanning process, and for changes in colour-temperature of the ambient light falling on the televised scene; and encoding the chroma signal onto a subcarrier according to the PAL (or other) specification. Along the way, the chroma signals are clamped to a reference level to prevent drift;

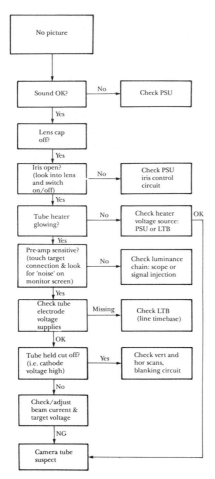

Figure 18.4 *TV camera basic fault-finding chart*

and gamma-corrected to maintain correct colour tracking throughout the brightness range of the scene.

When dealing with colour faults it is important to closely study the picture and any colours it *does* contain, comparing them with the original scene, which may be – as circumstances demand – a colour-bar chart, a plain card of red, green, blue or white, or a flesh-tone. Analyse as far as possible the defects and try to mentally assign them to the broad areas of the chroma processing section mentioned above. This can save much time in diagnosis. Some examples: a complete lack of colour should lead first to a check of the subcarrier output from the SSG, then the overall encoding and burst-insertion circuits, unless the picture is *black-and-green*, when the tube's beam focus is suspect. A loss of *one* of R or B is most likely to stem from a fault in the

colour-recovery section immediately downstream of the target-signal amplifier, or in the individual U and V encoding sections. An imbalance between R and B should first be investigated by checking the gain-control voltage inputs from the white-balance control section to the individual R and B amplifiers. A shading effect (graduation in brightness) in one or more colours across or down the screen suggests that the scanning fields are not linear, that the beam focus is not uniform, or that the shading controls need adjustment. And so on. A tube-type TV camera is a *very* 'analogue' device, for which the main diagnostic tool is an oscilloscope. Set the camera up to view the colour-pattern specified by the maker, and compare waveforms at key points with those given in the service manual.

Colour signal recovery

Some (mainly early) camera tubes like the Trinicon (Sony) and the Tri-Electrode vidicon have three separate target outputs for R, G and B, much simplifying fault-diagnosis and setting-up. More common is a single target connection, whose output contains basically Y (luminance) information, but also R and B information riding in a subcarrier generated by the action of the beam scanning across coloured stripe-filters buit into the tube's faceplate. Two basic systems are in use in 'domestic' cameras: phase discrimination and step-energy separation.

Phase discrimination system

The most common colour pick-up system uses a pick-up tube whose face is diagonally striped with gelatin-type colour filters of yellow and cyan arranged at opposing angles of about 25° to the vertical (Figure 18.5). The pitch of the stripes, and their width, is arranged so that their diamond intersections correspond to scanning-line spacing. Where yellow and cyan stripes intersect they combine to form a *green* filter diamond, so that alternate lines pick off the target signals corresponding to yellow/cyan, and to green/white. The angle of the stripes is such that as each new scanning line is traced the phase of its colour signal advances by 90°: this phase difference is the key to separation of colours corresponding to red ('anti-cyan') and blue ('anti-yellow').

Since the whole colour signal is based on stripe-repetition frequency (corresponding to about

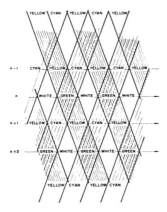

Figure 18.5 *Colour filter striping on the faceplate of an image-sensor. The 'n' numbers refer to line scans*

check the set-up and adjustments for scan amplitude and linearity, dynamic beam focus, and deflection yoke alignment.

Step-energy colour separation

An alternative method for extracting separate R and B outputs from a single target layer, sometimes used in domestic cameras, involves a pick-up tube whose faceplate has vertical stripes in the sequence green, cyan, clear, green, cyan, clear, over the whole picture area. As the line scan progresses the target output contains, in addition to the 'average' (Y) signal, a high frequency component in which information on (G), (G + B) and (R + G + B) colour components of the picture is contained in sequence (Figure 18.7). The stripe-repetition frequency here gives rise to an amplitude-modulated carrier at about 4.1MHz, which is filtered out from the target's signal and applied to a pair of opposite-polarity envelope detectors. Their outputs, in the absence of any colour in the picture, cancel out to give zero chroma output. When a coloured object appears in the picture the envelope outputs become unbalanced, giving rise to + R and − B signals at the detector's outputs; these are further processed into R and B and then R-Y and B-Y signals for encoding.

Again, the filter, detector and matrix circuits rarely give trouble, though in this type of circuit the field scan characteristics have little effect on colour rendition. Line scan amplitude and static beam focus have a large effect on colour amplitude however; and line scan linearity and dynamic focusing a large effect on colour shading, just as

3.9MHz) the chroma carrier is extracted from the preamplifier target output signal by a bandpass filter centred on this frequency. The separation process is carried out by a 90° phase shifter and a 1H delay line (Figure 18.6). Their outputs are applied to the colour-separator circuits, consisting of an adder to produce the B signal and a subtractor (invert-and-add) to produce the R signal. The separator and much of its associated circuitry is inside a chip, whose in/out signals can be traced by reference to the maker's block and circuit diagrams. Separator gain and phase adjustments for R and B are available for setting to minimum chroma output on a view of a white card. Seldom, however, do shortcomings in the separated R and B outputs stem from troubles in the recovery system itself. Before suspecting this area, exhaustively

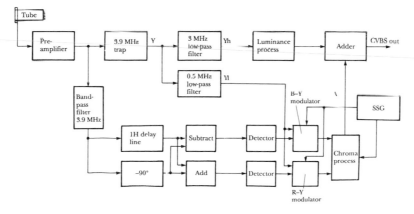

Figure 18.6 *Deriving a chroma signal from a diagonally-striped image sensor*

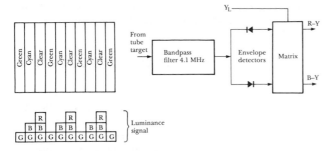

Figure 18.7 *'Step-energy' colour recovery system using a vertically-striped faceplate*

before. With either type of colour recovery system these 'beam' problems generally affect all colours, whereas the need for shading adjustment (described below) shows as *random* graduations for different colours down or across the picture. Another point, perhaps an obvious one, to bear in mind when fault-tracing in camera colour circuits is that a blue card before the lens gives rise to outputs in *both* B and R channels: positive and negative respectively. This polarity difference is not easily detectable when the signal is in carrier-modulated form!

Shading correction

Regardless of the form of the colour-separation method the signal effectively appears in R and B form in the separate output channels of the chroma separator. Reverting again to the block diagram of Figure 18.2, then, the next process is to balance out any shading of colours in the picture by applying carefully-trimmed 'equal and opposite' correction waveforms to the control inputs of a pair of VCAs (Voltage Controlled Attenuators) in each path, R and B. The correction waveforms are a complex mixture of sawtooth and parabola components derived from line and field scanning timebases by CR shaping networks. While the individual service manuals give full details, the basic process is to adjust parabola and tilt controls for zero colour (or colour-difference) signals (view on scope) throughout field and line periods while the camera is viewing an evenly-lit pure-white card. The need for adjustment here (and the efficacy of the job when done!) can be judged by inspection of the monitor picture (colour turned fully up) of a white card.

Colour highlight suppression

In principle, only the R and B (or R-Y, B-Y, V and U) signals are derived in the colour recovery and encoding circuits: the G (G-Y) signal comes basically via the Y (luminance) chain. This can raise a problem when the tube's target becomes saturated on a highlight spot, or when using the back-light control/manual iris override facility. The 'flattened-off' target signal loses its colour information, so that B and R disappear from the camera output signal to leave the highlights flooded with green. To prevent this a 'highlight chroma clipper' comes into operation to suppress *all* colour signals under these circumstances. Before adjusting or investigating this section for green highlight problems, check the working conditions of the pick-up tube, particularly beam current and target voltage settings.

Hue and saturation errors

The separate R and B channels are subject to differential gain control by the white-balance circuit whose 'error' outputs feed VCAs in each channel; incorrect hues, then, can be due to problems here – check the VCA control voltages against those quoted in the manual. VCA operation can be tested by applying a voltage from an external bias source to the control pins of the VCA chip.

Whereas white-balance faults affect the *amplitudes* of R and B signals differentially, hue errors arising from incorrect burst phase affect all colours, 'twisting' them around the colour vector-clock. Thus if the *mean* phase of the camera's burst signal is retarded, blue moves towards cyan, red towards magenta and green towards yellow. Incorrect burst *amplitude* affects colour saturation, via the receiver/monitor's a.c.c. circuits. In any case where reproduced saturation is incorrect, then, first check the amplitude of the burst signal emerging from the video output socket: it should be 0.3V, precisely equal to the height of the sync pulses. If it is, check the setting of any chroma gain preset. Adjustment of burst phase (and burst gain in those designs where the two are interdependent) is critical, calling for a vectorscope for checking. It is advised to leave burst control settings well alone unless the full test gear and colour charts specified in the service manual are available. Even when changing a pick-up tube, this adjustment is not normally required.

Hue errors are seldom introduced in the actual colour encoding process, though the complete loss of one of B-Y (U) or R-Y (V) signals may be regarded as a huge hue error! For this, check first for the presence of 4.43MHz subcarrier feeds to the balanced modulator section of the PAL encoder chip: a fixed phase for B-Y encoding and a line alternating carrier at ± 90° to it for R-Y encoding, as shown on the scope trace representation of Figure 18.8. If these are not correct, trace them back to the SSG chip outputs. If the carrier feeds are correct, check for baseband inputs to the chip, which, depending on design, may be in normal (R-Y, B-Y) or inverted (Y-R, Y-B) form; then for modulated subcarrier output(s) from the encoder chip. In some cameras, the modulated U and V carriers are combined inside the encoder chip, while in others (usually early designs) they come together in an *encode balance* preset. Lack of one or both modulated carrier outputs implicates the encoder chip, but check its pin voltages and peripheral components before condemning it.

Figure 18.8 *Relationship between the subcarrier feeds to the colour modulator/encoder section of a camera. The upper waveforms are both at 90° to the lower, and alternate at line rate in a PAL camera, giving rise to this trace on a 'scope*

The chroma-and-burst signal is combined with the luminance signal in an add stage for onward passage to the video output socket.

Record-only camcorders

Where no CVBS output or input is required, and the camcorder is designed for recording only, there is no need to generate a standard composite video signal. The luminance signal is fed direct to the recording f.m. modulator, and the chrominance signal is encoded directly onto a colour-under carrier suitable for the format in use. SW25 flip-flop pulses also enter this 'encoder', whose output is a standard colour-under carrier of the type described in Chapter 16. To service the camera section of this type of camcorder a special electronic jig (Y-C combiner) is required.

Camera performance evaluation

A colour-bar display, containing as it does *saturated* colours, is a poor test of the colour-rendering of a camera. Make judgement on a flesh-tone or a familiar picture having delicate pastel shades, having made sure that the monitor's grey scale is perfectly set up.

Solid-state image sensor

So far as servicing is concerned, CCD pick-up sensors are simpler than tube types, and there are far fewer adjustments involved. Guides to the theory of their operation are available from manufacturers and in text books. No high voltages, vacuum or heaters are required, and the timebases and magnetic scanning fields are replaced by drive pulses which step the light charges picked up by the silicon photodiodes down and across the device to be read out sequentially as a video signal.

In the block diagram of Figure 18.9 the CCD array is shown on the left, with its essential connections to the drive and video circuits. The lowest pin, no. 14, provides a bias for the photodiodes. No picture, or a blooming/flattening effect on whites in the picture should direct attention to the voltage here and the setting of V-sub control VR251. Pins 15–18 of the sensor are fed with vertical-rate (20ms) transfer pulses alternating between 'middle' (1V) and 'high' (13V) levels to step the picture-charges through the CCD cells towards the horizontal register at the bottom edge of the device. If these pulses from decoder/driver chip IC201 are faulty or missing the picture becomes blurry in a vertical direction.

As each line of picture charges reaches the bottom of the sensor chip they are stepped 'sideways' by the horizontal shift register, whose four-step sequence of drive pulses 0H1 to 0H4 is shown in the waveform diagram of Figure 18.10. A loss of all drive pulses (i.e. IC201 not working or no VD/HD trains from the SSG chip) will of course delete the picture altogether. Depending on camera design, absence of H drive pulses gives no picture or blurred horizontal streaks, while faulty pulse timing or amplitude render a picture with poor definition in a horizontal direction; for this check that the distinct

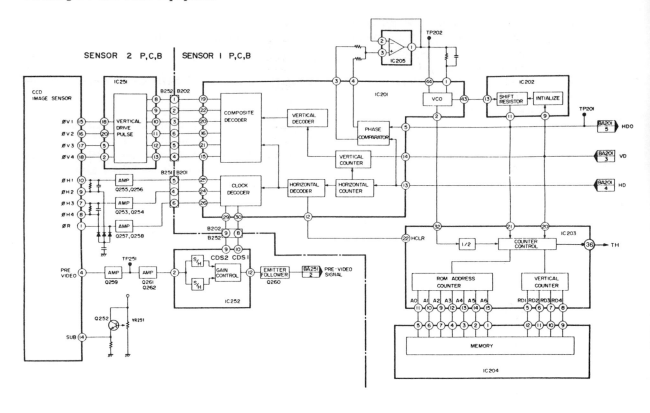

Figure 18.9 *Pulse drive system for a CCD image sensor (Panasonic)*

voltage levels (here 0–8V for H1/H3 and 3.5–11V for H2/H4) are present. So long as the H and V drive pulses are present and correct; that a gate reset pulse is present (here pin 1); and that supply voltages (i.e. + 12V or + 15V) are applied to the CCD image sensor, there should emerge from it (pin 4 in Figure 18.9) a recognizable video signal at several hundred millivolts amplitude, though at this stage much mixed with high-frequency 'hash'. After passing through a sample and hold gate (check sampling pulses, pins 29–30 of IC201 where necessary) the video signal leaves for processing.

The ROM memory chip IC204 and its attendant control chip IC203 are pre-programmed with the positions of element flaws in the CCD sensor, and substitute an adjacent 'good' pixel for each dud element in the matrix. This must be changed whenever the CCD image sensor is replaced – they come as a matched pair unless the new sensor is virtually flawless, when *no* ROM is supplied or fitted.

The face of the CCD sensor is striped with colour filters in the same way as described above for pick-up tubes, and similar techniques of bandpass filtering and colour separation are used to extract the colour signals; from CCD sensors the colour recovery process is more certain, since variations in 'scan amplitude' and 'spot size' cannot occur. The colour-balancing, encoding and burst-generation techniques are also similar to those used in late-model tube-type cameras. Similarly with the luminance chain: the same processes of clamping, gamma-correction, contour-enhancement etc. take place, as shown in the video process block diagram of Figure 18.11.

Although not connected directly with the use of CCD-type sensors, the circuit of Figure 18.9 shows that the 'scan generator' chip IC201 has its own master oscillator (pins 1, 2, 43, 44) which is phase-locked to the HD (horizontal) pulses from the master SSG by a PLL incorporating op-amp IC205. This feature is also found on cameras designed for *genlock* use. If the PLL comes out of lock, the pictures going onto tape (and viewed in E-E mode) are broken into lines or slipping sideways, because the output line sync pulses are sourced direct from the SSG chip, and the two are in disagreement. Check and set up the PLL associated with the drive-pulse decoder chip (scan-pulse generator in tube-type cameras).

Camera cable connections

Early portable video outfits consisted of separate camera and VCR, the two being linked by a multi-cored cable. A common problem here is open-circuit (usually intermittent) of one or more of its conductors, caused by flexing at one end or the other. In practice the stop/start (VCR plug pin 6) core is the most common offender. To help with ohmmeter tests of the conductors, Figure 18.12 shows the plug and socket connections and functions for the 10-to-13 pin system used on VHS 'separates'. Repair of these cables is not practical; once the fault has been proved, replace the entire cable assembly.

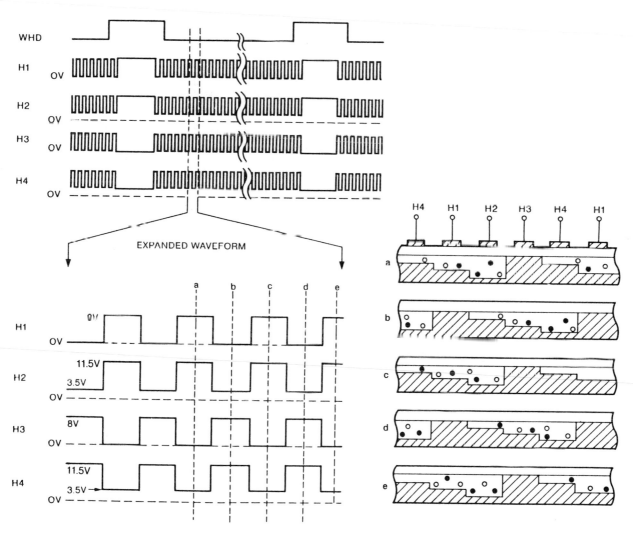

Figure 18.10   *CCD charge transfer timing by the 0H pulses of Figure 18.9 (Panasonic)*

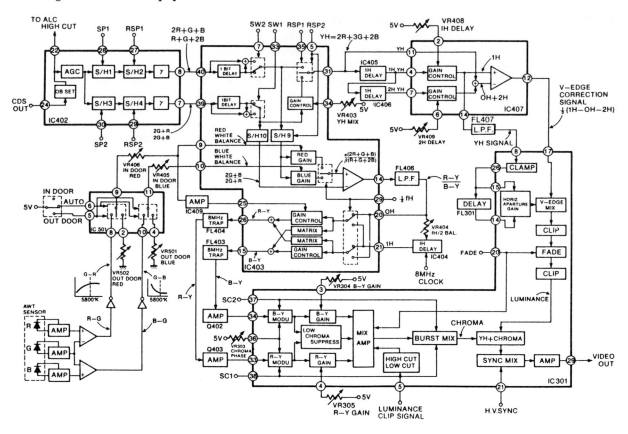

Figure 18.11 *Y-C process block diagram for CCD camera. The output of the image sensor is applied as 'CDS out' to IC402 pin 24 (Panasonic)*

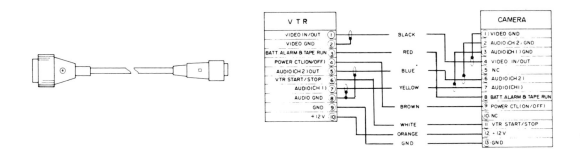

Figure 18.12 *Internal wiring of camera-to-VCR lead. The colours shown here may not be correct for all makes of lead (JVC)*

19

CLOCK, DISPLAY AND TIMER CIRCUITS

The section of a VCR dedicated to time-keeping and display is a trouble-free one, as a rule. Most faults reported by users are due to 'pilot error'; mains supply problems; or in the specific case of 'failure to make timed recordings' to deck-mechanical problems like slipping belts or idlers. Clock, timer and display-drive functions are all carried out by digital (often LSI) chips, in the roles of counters, memories and display-interfacing 'encoders'. In some designs a single LSI microprocessor chip is used for all these functions, which may also look after frequency-synthesized tuning, remote control and even some of the syscon operations described in previous chapters.

In this chapter we shall examine these sections from a fault-diagnosis viewpoint, concentrating mainly on homebase VCRs with fluorescent display panels.

Clock section

There are three main components of the electronic clock section of a VCR: a counter, counting from a reference source such as a crystal or 50Hz mains for timing reference; a display-driver section, which encodes the counter's output into pulse outputs to drive the grids and segments of the display tube; and the display tube itself which must, like a picture- or pick-up tube, be furnished with heater current, and an accelerating potential for its electron beam. If the clock reference input signal is missing, there will be no count, and in many cases no display light-up. If the clock gains or loses time, check first the timing reference frequency itself, which at some point (not always accessible) is brought down to 1Hz. Crystal reference oscillators typically run at 4.19MHz, and need very accurate setting with a counter. Where the reference is mains-derived, *small* timing errors are unlikely, so long as the machine is powered from the public

electricity supply: if the clock is 20% slow (takes 1 hour to count 50 min) check the state of any 50/60Hz pin on the counter chip. If the clock runs at half speed (counts 30min per real hour) check that any full-wave rectifier in the reference pulse feed is working properly and has 100Hz output. If these are in order, the counter IC itself is suspect once its supply voltage and relevant peripheral components have been checked OK.

Some counter/display systems can be set for 24 hr or AM/PM display at will: the programming is altered by making or breaking links on the IC pins. Details from manufacturers or service manuals.

Tape counter

The tape counter system is similar in operation to the clock, incrementing the reading by one for each pulse of the take-up reel sensor. Note that different makes and models of VCR do not *necessarily* count tape at the same rate – it depends how many sensor pulses are produced per revolution of the spool. Failure of the tape counter should lead to a check of the sensor pulses to the clock/counter/display chip: failure of the sensor itself will not be the cause if the deck continues to run. For counter-memory operation a 0000 reading triggers a coincidence-detector within the clock/display-drive chip, putting a stop signal into the syscon. If it doesn't work on FF and REW modes, check first that the 'memory' command is getting to the clock chip, then trace the progress of the stop-pulse from clock chip to syscon chip.

Timer faultfinding

Before we start examining the timer operation, it has to be said that in the experience of this author, perhaps 90% of timer fault complaints are not due to any malfunction in the timer electronics! Because the machine makes a 'cold-start' for unattended recordings, tuning drift (see Chapter 7) may be responsible for poor results, especially if they improve as the recording progresses; similarly, if the deck fails to complete its tape-loading cycle or the TU spool drive slips (see Chapters 12 and 13) reversion to stop mode takes place after a few seconds. These problems are most easily diagnosed by setting the timer for 2 minutes ahead, then carefully watching what happens at the appointed moment. Do it first thing in the morning when the VCR has been at rest in a cold workshop overnight.

Most other alleged timer faults are caused by user-error, and these are best discovered by watching the user set the timer. Typical errors are as follows: using a cassette with a safety-tab removed (many VCRs will flash an indicator or eject the cassette); incorrect day setting of the real-time clock or the timer programme; incorrect channel setting whereby a 'blank' channel is recorded; programming in 12hr/AM-PM language when a 24-hr clock system is used in the VCR; misinterpreting minute and hour/minute programming durations; accidentally erasing or altering times after completion of programming; and so on. Slavish obedience of the user's instruction book avoids these common pitfalls.

The block diagram of a typical timer chip set-up is given in Figure 19.1. The crystal X751 is the timekeeping reference source, replaced by 50Hz mains (pin 23) if the links between IC pins 14 and 15 are made. The 4MHz clock generator is a separate entity, used for 'housekeeping' within the chip. The IC's power source VCC (5.6V) enters on pin 32, backed up (in this case for about 5 minutes) during power cuts by large capacitor C751. When this occurs, the PF (power failure) detector opens all ports to minimize power consumption by swtiching off the indicators and display tube. The counter within the chip is controlled on parallel data lines for clock setting. The timer control section of the IC consists primarily of a RAM memory large enough to hold all the programmed data – start/stop times and channel numbers. When a coincidence between set time and real time is detected, data is passed to the tuning and syscon sections to select the required channel and start the deck in record mode.

Generally, the timer-programming data enters the timer control/memory chip via a key-matrix system, like those described in Chapter 13 and Figure 13.4. In case of problems, check first that the keyed-in data is registered on the display panel. If not, suspect the key-scan system, then the chip itself. If the data *is* displayed, but the timed operation is not carried out, check the link between timer and syscon sections before replacing ICs. Many machines have two record-start phases: one about ten seconds early, in which the tape is threaded up and record-pause mode entered; and a second, in which pause is released.

Some later designs of VCR have 'programmable' remote control systems, in which all the timer data is composed and displayed on the remote control

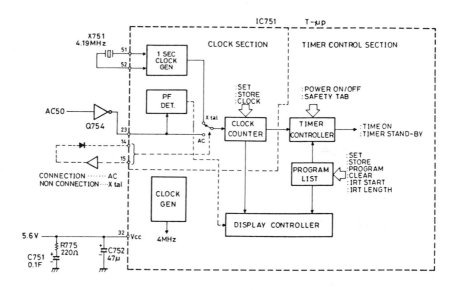

Figure 19.1 *Timer microprocessor chip arrangement (Hitachi)*

handset, then transferred in one 'squirt' to the memory bank within the clock/timer chip inside the VCR. For faultfinding, first check that the remote control is operating correctly for normal functions (Chapter 9) then use an oscilloscope to check for a burst of serial data into the VCR programme memory chip. Again, a useful indication of its entry and storage is given by the readout on the display panel.

Back-up systems

To bridge short interruptions in the public electricity supply a back-up system, in the form of a battery or large capacitor, is provided. It preserves real-time and programme memory data in the IC for a period (depending on design) from 5 minutes to several hours. This causes much misunderstanding amongst users, and careful reference to the VCR manufacturer's specification is essential to establish whether a fault really is present. If the back-up system is demonstrably not working or has short life, the battery or capacitor energy reservoir has probably deteriorated. If it (or its leadouts) are corroded or leaking electrolyte, replace it. If not, before condemning it, check that *all* isolation/charging diodes are neither leaky nor high-resistance, and that the charging current is high enough and of sufficient duration.

Clock roles

In faultfinding, it is essential that the operation of the various clocks are understood. The highly-stable *reference* clock has already been dealt with – it is the *timekeeper*. In addition, a *system* clock is required to step data-transfer and processing through and within the ICs – every microprocessor has one! These clocks do not have to be highly accurate or stable, and are usually timed by a ceramic resonator running at about 400kHz. Lack of system clock pulses will turn the display off altogether.

Fluorescent tube and driver

The display indicator is a cathode ray tube enclosing a vacuum. It has a heater, cathode, control grids (for want of a better phrase!) and anodes, the latter in the shapes of blocks, symbols and segments of

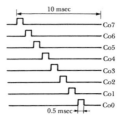

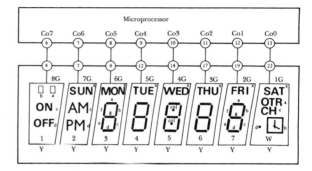

Figure 19.2 *Drive connections and grid pulse timings for a fluorescent display panel*

alpha-numerical characters, as shown in Figure 19.2. An accelerating potential of about 30V is required to light up the tube. In practice a – 30V line is generally used, with the anodes near ground potential, for convenience in driving the display from a chip.

The display area is divided into grids, here numbered 1G to 8G, and each embraces a group of segments and symbols. Each grid is continually pulsed on a sequential basis from the drive chip – see the time-related pulses in the diagram. These pulses alone do not light up anything, but make the segments in each grid sensitive to a second pulse train – that for the segments, labelled a–g and x in the diagram. Only when the pulses for grid and segment *coincide* in time will the designated area of phosphor light up. Thus to indicate ON in the display, the 'c' segment drive must receive a pulse in time-coincidence to the 8G grid pulse. A complete example is given in Figure 19.3, where the pulse train coincidences for a display of ON, SUN, 11–30AM are shown. These pulse trains can be checked with a double-beam oscilloscope. Trigger the sweep generator from one of the grid pulses and set the sweep period to cover the full sequence, i.e. 10ms in Figure 19.2. Selected segment and grid pins can now be examined simultaneously.

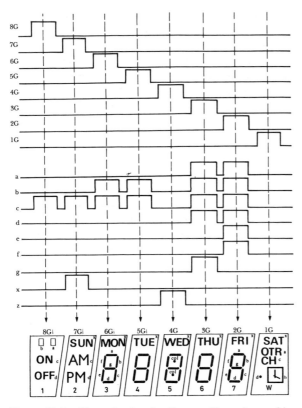

Figure 19.3 *Time-related pulse diagram illustrating strobing effect of fluorescent display drive*

Display faults

Perhaps the most common fault to arise is no light from the display at all. Check first that the machine is on and powered, then suspect a lack of tube operating potential (e.g. − 30V) due to a PSU fault, or (more often) a defunct DC-DC convertor module. If the operating voltage *is* present, check next that the tube is receiving heater voltage and passing current: typically 4Va.c. at its outermost lead-out pins. The front PC board, carrying user-operated keys, is vulnerable to hair-line cracks – and in some designs to liquid spillage, more of which later. With operating voltage and heater power present, the next check is with the oscilloscope for grid pulses. If they are present, firmly press a finger along the leadout pins: in these high-impedance circuits this introduces enough leakage to segment pins to light up sections of the display. If they don't light, confirm that segment pulses are also present, then check any 'display-enable' pins there may be. If these are in correct

state, the likelihood is that the display tube has lost its vacuum, sometimes betrayed by a cracked envelope or broken pip-seal at its rear or side.

If drive pulses are missing or incorrect, check that the drive chip has correct operating voltage and that its system and reference clock inputs are present. If so the chip itself is suspect. Odd faults encountered in practice in this area have been failure of 400kHz system-clock ceramic resonators; and failure of the shaping circuits which condition the 50Hz mains reference input to the counter chip.

More rarely, the symptom of unwanted segments alight may be encountered. If all segments are glowing very brightly indeed the likelihood is that the drive chip's operating voltage (normally + 5V, + 9V or + 12V) has greatly increased due to failure of a PSU regulator circuit. The tube generally survives the overload if it's not prolonged, but the drive chip could well be damaged. Sometimes all or some of the unselected display sections glow softly, see Figure 19.4. For this, a common cause is electrical leakage on the surface of the PC board in the area of the display panel or driver IC due to spillage of liquid into the machine – advice on dealing with this is given in Chapter 21. Other possible causes are the drive chip and associated zener diodes, etc. – anything that reduces the cut-off bias of the fluorescent tube. Odd incorrect sections alight are most likely to be due to a faulty driver chip, but check for electrical leakage between the pin connections of the IC and the display tube.

For a dim display, check that the accelerating voltage is up to the level quoted in the manual (26Vor 30V as a rule) then that the tube's heater voltage is adequate, and that the duty-cycle of the

Figure 19.4 *Spurious illumination of display sections*

strobe pulses is correct. One or other of the latter two are sometimes controlled by a dimmer switch, or by the on/standby control: check the setting and condition of these where relevant. If they are in order the display tube itself is suspect for low-emission or tired phosphors. One symptom of a worn display tube is uneven illumination of the segments.

If the display tends to flicker excessively, check for synchronization between the frequency of the heater voltage and the grid drive pulses – some VCR designs have special arrangements for this.

Clock reset

Occasionally a machine may be received for service with the complaint that the clock intermittently flashes. Usually this means that the display shows a flashing 12:00, 00:00, 88:88 etc. due to the control microprocessor being reset. Apart from the obvious cause of a power cut, this may be caused by dry joints in the PSU section; by a fault in the reset circuit itself; or by lightning, static or interference pulses on the mains supply. In 'electrically noisy' districts, the latter cause can often be mitigated by using a 'power-cleaner' device, as designed for use with computers. Check too for latchety mains plugs, sockets and connections, which can also, in rare cases, cause the clock to run fast.

LCD indicators

The liquid-crystal display panels used on battery-operated portable equipment, with no filament, vacuum or high accelerating voltage, are even more reliable than the fluorescent type. The LCD display, like its thermionic counterpart, is divided into grids and segments, with separate lead-outs for each. The drive voltages come from an encoder IC as before, but are generally in a.c. form – a squarewave of about 5–9V p-p with a repetition frequency of 50–100Hz. For no-display faults check the supply voltage and clock oscillator of the display-drive chip before suspecting the IC itself.

20

DEALING WITH INTERMITTENT FAULTS

Of all the problems which can arise in TV and video equipment, intermittent and spasmodic ones are the most difficult and frustrating ones to trace and cure. In general, intermittent symptoms stem from the same causes, and respond to the same diagnostic procedures, as given in preceding chapters of this book, so here we shall concentrate mainly on the difficulties in diagnosing transient faults; methods of provoking them; and effective soak-testing. It is important to get as much information as possible from the user regarding the symptoms and circumstances of the absent fault. Since users are notoriously vague about these, if possible get a sample tape of any recording fault, or where necessary in the case of video-playback or TV-set troubles, simulate fault conditions as examples for a yes/no reaction by the user. For 'lines all over the screen' try breaking the line hold, unlocking the colour sync or upsetting the tracking as appropriate; for 'wrong colours' take out R, G and B one at a time on a TV, or upset the white balance on a camera; for 'no picture' pull out the aerial plug, switch to AV, etc. in the presence of the user, who may thus be able to define the fault symptom more clearly.

When an intermittent fault *does* perk up, with or without provocation, try not to lose it! First closely study the symptom, relating it to possible causes and sections of the equipment; then attempt to take meter and scope readings – as many as possible 'on tiptoe' while the fault is present. Do not, whatever you do, try to make the fault go away again by mechanical or thermal attacks!

Types of intermittent fault

Whether in mechanical or electrical sections, intermittent faults fall into one of four categories, or combinations of them.

Physical faults

These are the most common, and can usually be provoked by 'mechanical' means. Printed-circuit board trouble is perhaps top of the list, in the form of hair-line cracks and dry-joints. Where a hair-line crack is responsible, it is often more visible from the *component* side of the board (Figure 20.1). Close and careful inspection of both sides of the PCB at the outset is recommended: look particularly at the areas of heavy board-mounted components, board fixing screws and any narrow 'neck' sections of PCB. Most hair-line cracks can be detected by flexing the PCB as shown in Figure 20.2, where pressure on the component side of the panel induces the fault, but pressure on the print side relieves it. Tracing along a print run with a voltmeter (equipment switched on) or ohmmeter (equipment switched off) shows the point of break, which can be almost invisible to the eye. For this a needle-sharp probe point is helpful to pierce the lacquer coating; an audible bleep continuity test, as

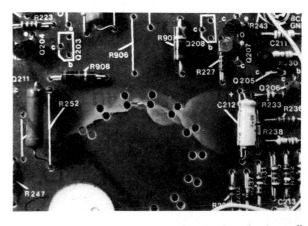

Figure 20.1 *PCB surface crack, showing the 'spread' effect*

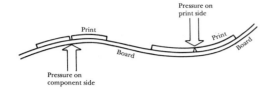

Figure 20.2 *Probing for hairline cracks on a PC board*

featured on many digital DMMs, is useful in facilitating 'eyes-off' testing.

Dry-joints are most easily found by gently tapping and flexing PCBs and components – in difficult cases, as soon as the fault has been provoked use a voltmeter or scope to trace the open connection. Bear in mind that even quite small voltages and currents give rise to tiny arcs at the trouble-spot, often visible in a completely blacked-out workshop once the eye is accustomed to the dark. With experience, suspect joints can be thus detected on inspection. Figure 20.3 is a close-up shot of two very bad ones! Dry joints are most likely to crop up where a large mechanical, thermal or electrical load is present: at the tags of heavy components like LOPTs, large electrolytic capacitors etc; at scan-coil and PSU connections; on heavy wire-wound resistor lead-outs (Figure 20.3); and at mains-switch PCB tags, where arcing is particularly dangerous.

Figure 20.3 *Two adjacent dry joints are visible here; the large joint left of centre is also suspect*

Wherever electrical contact depends on spring pressure (plug/sockets, switches, potentiometer wipers, motor brushes, CRT pins, etc.) there is a risk of unreliability due to tarnishing, dirt or loss of spring pressure with ageing. These vulnerable points are the first to check where tapping and flexing shows the fault to be a physical one. Unless the trouble-spot is obvious try to *prove* the discontinuity for a certain diagnosis.

Many other physical causes of intermittent problems can be found by close examination of the set's interior under strong light, if necessary with a magnifying glass. Look for discoloured or cracked resistors, pinched wires, chemical leakage from capacitors or batteries, desposits from liquid spillage, foreign metallic bodies, loose earthing screws, burn marks, etc.

Thermal faults

Though it's not always obvious, many intermittent faults are temperature-dependent, with the symptom only arising from cold switch-on, or when the equipment has thoroughly warmed up. Very often, thermal expansion/contraction is at the root of this, though semiconductors can develop internal faults at temperature extremes. Fortunately, the means of detecting these are readily available in the forms of a hair-dryer and an aerosol freezer spray, with which suspect components can be temperature-cycled at will: always bear in mind, though, that the characteristics of semiconductors, particularly diodes, zeners and transistors, are dependent on temperature, and that this may be balanced in some circuit designs by 'equal and opposite' compensation from another component sharing the same environment. Unless the two receive equal treatment in terms of heat-and-freeze, the effects may be misleading. In a crowded PCB layout, it may be necessary to isolate the suspect from its neighbours by fitting a piece of suitable gauge sleeving over it, or re-mounting it on the print side of the panel before heating or cooling it. To heat very small areas, use the hot-air stream above the bit of a soldering iron.

Thermal problems can also arise in mechanical components. A motor whose bearings are stiff may possibly run when cold, but reach siezure point in higher temperatures. Conversely, any mover running in old, dried-up grease may have too much friction to move when cold, but if physically pushed, or allowed to reach a higher temperature, will operate. The performance of friction surfaces of rubber (idlers, drive belts) if in poor condition may also be temperature-dependent.

Random faults

These are the most difficult to trace, since they cannot be provoked by mechanical or thermal means. The most important thing is to glean as much information as possible about the fault while it *is* present – by observation and by instruments. Then progressively narrow down the field of search by setting diagnostic instruments on the basis of previous readings. Further techniques are described later in this chapter.

Parametric faults

Many intermittent problems which appear to fall into one of the three categories described above are due to permanently-incorrect settings or conditions within the equipment, needing only a slight change in operating, signal or thermal conditions to cause trouble. Typical examples are a servo loop or line sync circuit which is running just within its hold-in range and needing only a hiccup in control-track continuity or a change of channel to lose lock; a too-high setting of regulated PSU output on the point of triggering an over-voltage protector; an inefficient sync separator which invokes field rolling on certain scenes only; a maladjusted VCR inlet guide giving rise to a shortcoming in envelope signal such that the replayed field sync pulse is borderline; an incorrectly-timed burst-gating pulse which permits *signal* chroma to influence the subcarrier PLL, depending on picture colours; high-resistance or O/C PSU start-up resistor, making switch-on from cold a hit-and-miss affair; unpredictable tape tracking due to the ribbon not sitting firmly down on the lower-drum rabbet; leakage across PC board surfaces due to liquid having been spilt onto them; a whole chapter could be filled with examples of this sort. Suffice it now to say that wherever an intermittent fault is present it is well worth while to closely inspect the affected area, to check waveform amplitudes and supply-voltage levels, to ensure that mechanical and electronic presets are correctly set, and to check for the effects of wear, drift and ageing.

Positive diagnoses

In dealing with intermittent problems it is very easy to fall into the trap of replacing one or more suspect components (without having made a real diagnosis) then releasing the set after a trouble-free test run. In many cases this cures the trouble, but there is a strong risk of the job 'bouncing' with the original fault still present, and the customer justifiably demanding that the job be re-worked at no further charge or a reduced charge.

To avoid this embarrassment (and for greater job-satisfaction!) it is better to adopt a *positive* approach to intermittent faults, in which definite proof of the diagnosis is obtained. This can be achieved by inter-substitution of parts, where the suspect component is swapped with a similar one – in the same set or another – until the fault perks up in the previously-good channel to give a 'cast-iron' diagnosis. This is most easily done where the set contains identical stages (i.e. stereo, RGB, logic and 'array' chips, etc.) but can with thought be applied in many other situations.

Similarly, some meter checks can give a positive indication of a faulty component: a voltage appearing across the 'closed' contacts of a switch or plug/socket; the presence of (e.g.) 12V across a 7.5V zener diode; a d.c. current flowing in a capacitor; zero current in a load resistor when a voltage is present across it. Even if such readings last only a second or two while the fault is present they are enough to positively clinch the diagnosis. Fig. 20.4 shows some further examples. In Figure 20.4(a) if the voltmeter reading rises above 800mV (i.e. during heat/freeze cycles) the transistor is definitely faulty. Figure 20.4(b) shows two meters temporarily

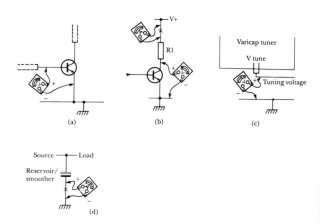

Figure 20.4 *Meter hook-ups for continuous testing for intermittent faults. In diagram (c) a microammeter in series with the tuning line can also be useful; in (d) beware of overloading the meter during switch-on and switch-off charge/discharge surges. Procedures are described in the text.*

hooked to a video output stage. If when the screen floods with light the current sinks to zero along with the voltage reading the load resistor R1 is going O/C. In Figure 20.4(c) a high-impedance meter is left hooked onto the Vtune pin of a varicap tuner. If the voltage here *falls* when the tuner drifts up-band the tuner itself is irrefutably responsible for the fault. Figure 20.4(d) suggests a method of checking a decoupling or smoothing electrolytic capacitor with an a.c. millivoltmeter: if the measured current decreases when hum/ripple symptoms appear the capacitor can be condemned with confidence; if it increases, however, the capacitor is doing its job and the fault is elsewhere, perhaps due to excessive current demand in the load. Again, these examples are by no means exhaustive, but they may suggest other ways of achieving positive diagnoses in the limited time during which an intermittent fault is present.

Tests for intermittent faults

In stubborn cases, where the fault cannot be brought on by mechanical or thermal provocation, and where the operating conditions are found to be within tolerance, the set must be left to run with test equipment permanently hooked onto key test points. It's not practical to tie up the main bench test equipment, but most workshops contain out-dated test meters, old scopes with low-emission tubes and similar discarded gear which is usually adequate for this purpose. Following the advice given in the appropriate chapter of this book, isolate the fault to one stage of the equipment (sequence: voltage supply, signal in, signal out, operating conditions) then set up tests for individual components.

Sometimes the fault's duration is so short that it's difficult to attend and read-off the instruments' indications in the time available. Such solutions as a storage oscilloscope, a strip-chart recorder or a camcorder watching the *ensemble* are possible, but are not usually practical or economic. The arrangement shown in Figure 20.5 may help in some circumstances. A suitable relay is wired so that its solenoid current flows through its own normally-open contact. Connect to a supply line under test and latch in the armature by hand. If the supply voltage momentarily collapses the relay drops out and stays out to record the fact. A simple one-transistor amplifier can be added to make the device more sensitive if required. In the same vein,

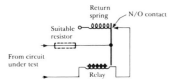

Figure 20.5 *This latched-on relay will trip out and stay out if its feed momentarily disappears*

Figure 20.6 suggests another latching circuit, this time using a small thyristor to permanently bring on a bulb or LED whenever a pulse or voltage level exceeds a pre-set level: use of a zener diode or resistive divider in the gate circuit adapts the device for different circumstances. This and similar 'home-made' trap circuits do not take long to devise and assemble, and once made can be kept on hand for use as required.

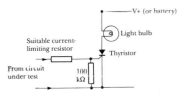

Figure 20.6 *The indicating light bulb comes on and stays on to record a transient pulse at the thyristor's gate. Almost any type of thyristor is suitable for this test-jig.*

Short component life

Also falling into the intermittent category, perhaps, are situations where a newly-replaced component fails again after a relatively short life. A common cause of this is the use of incorrect or unsuitable replacements. Ensure that belts, idlers, fuses, power transistors, triplers, etc. are correct type, correctly fitted and preferably supplied by the original setmaker: the 'equivalent' types given in data books for power-switching and line-deflection transistors and diodes are not always reliable replacements. Be wary, too, when replacing capacitors which operate with high ripple currents and/or 'high' frequencies in power-handling circuits. The normally-quoted ratings of capacitance and working voltage are not sufficient in themselves to describe these – compare the two capacitors in Figure 20.7, which have superficially similar

Figure 20.7 *These components have similar capacitance and working-voltage ratings. The one on the left will quickly overheat and fail if asked to do the job of the other!*

characteristics. Physical size is a useful rule-of-thumb guide here.

Electrical breakdown, mainly of semiconductors, can be repeatedly caused by intermittent flashover in the picture-tube, EHT rectifier assembly or focus-control section. Check high-voltage settings, and where necessary consult the manufacturer's technical department. Some designs are more prone to this than others.

Repeated failure of (correct-type!) line output transistors is generally due to excessive dissipation or punch-through by too high a collector voltage pulse during flyback. In the former case check that the transistor is properly heat-sinked, and that the base drive current is adequate and has fast rise and fall times. If so, suspect excessive loading in the collector circuit due to S/C turns in the LOPT, etc. The current drawn by the stage is a good guide to this. Where it is suspected that excessive flyback voltage is responsible the first suspect is the tuning capacitor, C1 in Figure 4.5, p. 38. Intermittent failure of the efficiency diode or the EHT rectifier are also possible (though less likely) causes of this.

The above remarks largely apply also to the main switching transistor in chopper power supplies, which works under similar conditions to line output transistors.

Mechanical components in VCRs can also sometimes suffer from short-life problems. A video head drum which wears prematurely has probably suffered from too-frequent use of a cleaning tape (check with user) or too high a back-tension in the tape, which will itself wear quickly under these circumstances. The need to replace VCR belts or idlers at frequent intervals suggests that excessive mechanical loading (brakes not coming off, lack of lubrication?) or oil-contamination is responsible.

Intermittent fuse blowing

A difficult fault to deal with is spasmodic failure of fuses, which may last hours, days or weeks before blowing. The first action should be to ensure that the correct rating and type of fuse is in use. Sometimes manufacturers change fuse specifications as a 'field modification'. Next check that some permanent fault is not present, pulling excessive current through the fuse; and that its holder is in good condition. Examination of the failed fuse tells much about its fate, as described on p. 30 of Chapter 3.

Quiz the user about any events (flashover, arcing, loss of line scan etc.) at the moment of failure which may assist the diagnosis. Common causes of intermittent rupture of mains fuses are a faulty filter capacitor (fitted across the mains input downstream of the fuse); and operation (where fitted) of a *crowbar* circuit, whose effect is to short-circuit the mains in the event of over-voltage or over-current situations. The crowbar may be firing legitimately (check PSU output voltage for creep-up during a soak test) but just as often electrical leakage in a zener diode, thyristor or other semiconductor lies at the root of the problem – see Chapter 3.

Other possibilities for random fuse-blowing are faulty PSU semiconductors; dry joints/arcing in the PSU or LTB; S/C turns in mains transformers; stalling motors in VCRs; and failure of soft-start circuits (check C-R charging elements) in switch-mode power supplies, leading to a high surge of current at switch on. Latchety connections at mains or battery plugs can also lead to sporadic fuse failure.

Experience in the field has led some setmakers to introduce circuit modifications or revise adjustment procedures to overcome problems of intermittent fuse blowing. Check with the technical department of the manufacturer concerned.

Vulnerable components

Some types of components rarely develop intermittent faults, while others are more prone to them. Where the fault cannot be narrowed down to less than one stage, and it becomes necessary to replace components 'on spec', experience suggests that the following classes are suspect for intermittent and thermal problems, given in order of likelihood: semiconductor devices, tantalum capacitors (especially surface-mounting types), aluminium electrolytic capacitors, skeleton pre-sets, polystyrene capacitors, low-voltage ceramic capacitors. Other types of components can of course give rise to intermittent problems.

Examples from practice

Several examples of intermittent fault causes have already been given in this chapter. Others, based on experience, are as follows, representing a 'top ten' of those not previously mentioned:

1 *Distorted sound after run*: warping loudspeaker cone, incorrect quiescent current setting, faulty output transistors or IC.
2 *Tuning drift*: faulty two-leg 33V stabilizer IC, a.f.c. left switched off by user.
3 *Focus changes*: corrosion at tube pin, corona discharge at focus spark-gap, leakage across tube base-panel surface.
4 *Loss of field sync*: low-amplitude sync pulses due to crushing in the i.f. amplifier. The a.g.c. circuit is often responsible.
5 *Intermittent colour from VCR*: crystal oscillators drifting out of tolerance, PLLs need setting up.
6 *Spasmodic VCR deck shutdown*: insufficient take-up torque.
7 *Erratic VCR servo operation*: weak/borderline PG or FG pulses due to incorrect magnet spacing etc.
8 *Sporadic loss of VCR replay servo lock or muting*: worn, dirty or wrongly-positioned control-track head.
9 *Reversion to standby/stop/channel 1*: poor contacts in mains plug, adaptor, switch or PSU.
10 *Occasional failure of VCR to start from rest*: 'dead-spot' on commutator of d.c. motor used for cassette/tape loading and reel drive. Some older VCRs use brush motors for capstan and head-drum drive.

21

REPAIR TECHNIQUES

Throughout the many chapters of this book we have looked at diagnostic and fault-finding procedures. When the faulty component has been found its removal and replacement is generally a simple matter of desoldering the old one and soldering in the replacement with a hot clean soldering iron and flux-cored 60/40 solder of 18swg for ordinary work, 22swg for tiny components and surface-mounted devices. Bear in mind the advice given in appendices 1 and 2 when selecting and fitting replacement parts.

Virtually all TV and video equipment is based on the concept of printed circuit boards in plastic cabinets, and this chapter will concentrate mainly on the practicalities of dealing with these.

PCB work

Of the several methods of de-soldering joints to remove components from a printed-circuit board, the author's experience over a long period is that flux-impregnated braid is the most satisfactory. As shown in Figure 21.1 it is placed between the soldered joint and the iron to suck up the solder by capillary action, particularly effective with multi-legged components.

Perhaps the most common PCB fault is a crack, breaking the continuity of the printed conductors. Check carefully the extent of the crack (not always easy to see!) and where necessary drill a small hole at its end to stop it spreading further. Use a new, sharp watchmaker's mini-screwdriver or similar tool to scrape the surface enamel off the affected print tracks for say 5mm on each side of the break. The mini-screwdriver, if pressed hard and twisted, can also be used to drill holes in the PCB. Tin the exposed part of the tracks and bridge the gap with a soldered-on wire bent to the shape of the track as shown in the drawing of Figure 21.2. For most purposes 5A fuse wire is ideal for this, though something thicker (e.g. tinned copper wire down to 12swg) may be required to reinforce and support a *broken* PCB. In non-stressed areas of the PCB it is sufficient to lay a heavy trail of solder over the scraped track in the area of the crack. Where the PCB pattern is very fine the bridging of cracks is a delicate and painstaking job, which may be aided by using a magnifying glass – an illuminated bench magnifier to work under, and a ×8 watchmaker's eyeglass for subsequent minute inspection of the job.

Incidences of heat-damaged PCBs have reduced over the years due to cooler-running, low-dissipation

Figure 21.1   *Desoldering braid in use*

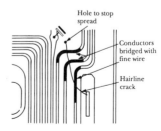

Figure 21.2   *Repairing a cracked PCB with fuse wire. It's easiest to cut the bridge-wire after it has been soldered in place*

designs and the influence of BEAB requirements. If burning or carbonization of the PCB base material *is* encountered, all the affected material must be cut away, since it is conductive – file and cut back to clean material in all directions, bridging the holes thus made with tinned copper wire of suitable gauge, and making loops where necessary to connect and support components. In severe cases a suitably-shaped patch may have to be fashioned from SRBP or paxolin and glued into place on the holed board, with the print tracks replaced by wire. Weigh up the labour time and cost involved, and the safety-factor in subsequent use, before embarking on such a repair!

Surface-mounted devices

Cameras, camcorders and other miniaturized equipment make use of surface-mounted devices (SMDs), whose removal (and replacement) is more difficult than that of conventional components. The main problem is in completely breaking the bond between the component's lead-outs and the PCB without damaging the latter. For a two- or three-connection component (resistor, capacitor, diode, transistor) use desoldering braid to suck all the solder from the connections, then twist the component with a pair of pliers, when it should come free. Even if the component is not destroyed in removal do not use it again.

For multi-pinned components (usually ICs) several methods of removal are possible. Lesser-used ones are the use of a frame-shaped iron bit (or bit-extension) to heat all pins simultaneously; and application of a hot-air blower to achieve the same result, when the component can be lifted clear of the board. These methods do not seem as practical or satisfactory in actual use, however, as individual detachment of legs. Figure 21.3 suggests a process which is clean and usually successful. Remove as much solder as possible with braid, so that the lead-out feet are as bare and dry as you can make them. Now thread a wire (e.g. 5A fuse wire) under the pins along an edge of the IC package, and anchor it at the far end on a PCB land. Pull the free end of the wire (keeping it as close and parallel to the board as possible) as each IC pin is heated in turn with a hot dry soldering iron. Each pin is thus released in turn as the wire passes under it. Repeat for the other ICP edges until all lead-outs have been freed.

If the method outlined above does not work, heat

Figure 21.3 *Removing a surface-mounted IC in 'cheese-wire' fashion. Here the wire is anchored to a solder blob to the 'south-west' of the IC package*

each pin in turn while levering it away from the board with a small metal hook or screwdriver. The alternative method of cutting away the pins for separate removal afterwards is only recommended if you are *very* careful – one slip of the cutting tool can severely damage the board and other components.

When replacing a multi-legged SM component, its correct positioning on the PCB is crucial for success. Stick a pin-head size blob of blu-tack or similar under the centre of the device (Figure 21.4) to hold it in place, then exactly centre it on the connection pads of the PC board. First solder opposite corners to anchor it, then solder each pin in turn. Most SM ICPs come ready-tinned, and so long as the PCB pads have been lightly tinned first, there is usually no need to apply extra solder when attaching the new ICP. This avoids the risk of introducing shorts with solder blobs. Closely examine all the joints with a ×8 eyepiece magnifier before switching on!

SM components should be dealt with using a very fine (25W or less) soldering iron whose bit is

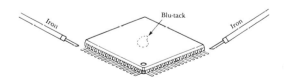

Figure 21.4 *Initial placement of an SM component is crucial. Anchor an IC with a central adhesive blob, then tack opposite corners down*

about 1.5mm in diameter. A bigger iron can be adapted by making a bit extension from tinned-copper wire of about 18swg – wrap the wire around the bit, leaving a stub about 10mm projecting forward as shown in Figure 21.5.

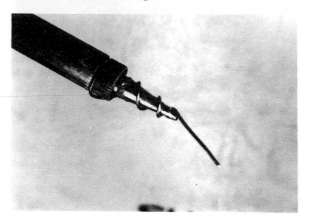

Figure 21.5 *Fine-gauge solder bit improvised from tinned copper wire*

Because many two-leg SM components are too small to have their value and ratings marked on them, keep them in the packing until the last moment before fitting them. The same applies to ICs, many of which are static-sensitive C-MOS types, protected in transit by conductive packing.

Liquid spillage

A problem which appears to be getting more frequent is that of users spilling liquid into electronic equipment. The most common entrants are drinks, alcholic and otherwise; water from flower-vases and pot-plants; and worst of all, pet urine. Especially when electrolysis sets in on the surface of PCBs and components, all these liquids are very corrosive. At the outset of the repair, very carefully check the severity and extent of the contamination before estimating or embarking on the job. If the liquid has entered a motor, transformer or tape-head assembly, it's likely that the copper windings will be attacked, 'writing-off' the component, which can be expensive; if the upper or lower drum peripheral surfaces are corroded they will have to be replaced; severely affected PCBs may be beyond repair, and complete board assemblies are seldom available as spare parts. Very often the extent of the damage is not obvious or visible, and even after apparently successful repairs

there is a heavy risk of 'bounce-backs' as the corrosion eats into metallic surfaces and spreads its conductive influence. Many severe cases have to be written off if a realistic estimate for materials and labour time is made.

It is best to replace any friction surfaces (belts, idlers, tyres, pinch roller) the liquid has touched, and treat metal surfaces with solvent-soaked cloth or cotton buds. Lightly lubricate sliding and bearing surfaces before reassembly, and thoroughly clean the tape path and heads.

All affected printed-circuit boards should be removed from the machine altogether if possible, then all components in the affected area whose bodies are in contact with the board surface should be completely removed. Lift off the bulk of the deposits and corrosive salts by dabbing with a solvent-moistened cloth or bud, then use a small paint brush to flush the board with surgical- or methylated-spirit or isopropyl alcohol. Do not use paraffin, petrol, turpentine or similar solvents. Use the paint brush, and a stiff tooth-brush where necessary, to scrub the board and components, rinsing often to remove contaminents. Do the same with the ICs, holders, plugs, sockets and other components removed at the outset, dunking and scrubbing them with solvent. Finally dry all with a cold blast from a hair dryer.

Examine all components, and particularly their leadouts, for signs of corrosion or weakness, replacing as necessary. It's a good idea to check capacitors, diodes and transistors individually for electrical leakage before re-fitting them, discarding any which are doubtful in the slightest. Switches, coils, mini-presets and similar vulnerable parts should be replaced anyway if they have suffered liquid ingress. Re-make all suspect joints and check the state of any leads which are soldered or crimped to the affected area: the liquid travels up them (inside the sleeving) by capillary action, and causes green or blue corrosion.

If when all is assembled the electrical operation is still incorrect it's likely that leakage and corrosion is still present somewhere, especially if the fault defies normal analysis: the effects of random leakage paths between pins, tracks and adjacent components can be very strange.

Cabinet and plastic repairs

In general it is better to replace rather than repair cabinet and plastic parts, especially since repairs may infringe safety standards and BEAB requirements. In many cases, however, cabinet parts are not available as spares for long after production has ceased, and in others parts are hard or impossible to get due to the age or obscure make of the equipment. Sometimes expense precludes cabinet or large-component replacement.

A glue-gun, using hot-melt adhesive sticks and worked by gas or electricity, is useful for all types of cabinet repair work: mending cracked cabinets (get the glue well into the crack and build up a thick layer below); re-fixing broken switch, facia and push-button mountings; restoring pushed-out aerial socket mountings; anchoring broken back-securing lugs; and so on.

The thermoplastics used are of course easily melted with a soldering iron, and this can sometimes be used to advantage: 'plastic welding' with a hot iron or gun can be used to build up and repair broken sections, but make sure the work area is well ventilated. Where a piece has broken off a plastic member, a repair of sorts can sometimes be effected by melting a wire stub into place. Cut a short length of tinned copper wire and steer it into place with pliers at the outer end, while heating it with an iron near the point of entry. Withdraw the heat quickly for a good effective bond. This technique can be useful for restoring broken 'swing-spigots' on cassette flaps and tuner/control doors, though again replacement is far better if the parts are available.

Cobbling

Most 'repairs' to TV and video equipment are not repairs at all – rather the replacement of defective parts by new ones. Apart from the cabinet and plastics repairs outlined above, however, sometimes it's necessary to repair rather than replace major components on economic grounds, or where a replacement is simply not available.

A common example of the former is where the centre-pin or outer leaves are broken off a VCR r.f. booster/modulator aerial input or output socket. These somewhat fragile connectors, in some designs, are crimped to the body of the r.f. module and at first sight appear impossible to remove or replace. Before condemning the module try this: first desolder the centre-pin and outer lugs from the PCB, ensuring that they are entirely free. Now grip the remains of the body of the connector in a heavy pair of pliers and twist and rock it, levering if necessary

from below with a screwdriver. With care the socket can be worked loose and removed without damaging the modulator/booster beyond bending its case. Once out, the plug or socket can be replaced with a commercially available surface-mounting type, soldered to the module body; in some cases it's more practical to remove the fixing wings from the replacement before sweating it on. Use one or more heavy soldering irons or guns for this, and fit a spare co-ax plug while doing it, to act as heat sink. This lessens the risk of melting the plastic insulating support inside the socket.

Similar economic necessity can arise where a LOPT or high-voltage component (or associated sealed-in lead) develops a pinhole through which violent sparks jump to adjacent conductors. Remove the component for easy handling, then roughen the pinhole and surrounding area with a file or scraper. Now clean it with a solvent and allow to dry. Apply a large quantity of newly-mixed quick-set epoxy resin, forming it into a big 'teardrop' over the pinhole by turning and manipulating the component as the epoxy sealant dries, aided if necessary by hot air from a hairdryer. Two applications may be necessary to build up a thick layer for good sealing and insulation.

Bench organization

For a busy TV workshop with high throughput, an effective mode of working is to use large double-decker trolleys to transport, repair and test the sets on. The 'bench' can then consist of a narrow peninsula against which the trolley is pushed; when repair is complete the trolley, complete with TV, can be wheeled away for soak test and replaced by another: indeed the set need not leave its trolley throughout its stay in the workshop. Similarly, the mirror can be mounted on horizontal pivots in a frame on castors for easy manipulation. The problem of accommodating large circuit diagrams and service manuals can be solved by a large 'reading board' on horizontal pivots, with a retaining bar at the bottom. This is easily improvised out of a wheeled TV trolley and a pair of uprights; the 'tray' section of the trolley forms a useful stowage for the tool-box. This and the mirror are sketched in Figure 21.6.

For a right-handed technician, a practical arrangement is to have the test equipment on the bench and shelves to his right-hand side; fit aerial sockets and mains-isolated 13A mains sockets, and

Figure 21.6 *Workshop accessories 'on skates' are convenient and versatile. Shared test equipment (high-spec scope, etc.) can also with advantage be trolley-mounted*

a 'safebloc' connector, at the outer (left-hand side) end of the peninsula in front of the technician. This saves having wires draped across the field of operation. For servicing VCRs, the conventional 'table-type' bench is probably best, with the under-repair VCR flanked by the oscilloscope on one side and a 14″ or 16″ TV/monitor on the other.

The bench lamp should ideally be of the circular fluorescent type with central magnifying glass. Filament lamps are hot, unreliable and much less convenient! To supplement this broad spread of light, a useful additional tool is a miniature inspection torch with flexible swan-neck: it can brightly illuminate small inaccessible areas, and some designs have interchangeable tips, so that when a screw is lost in the works it can be retrieved with a magnet screwed in place of the bulb.

Mention of screws leads on to the stowage of them while repairs are in progress. The ideal is a multi-compartment tray, with each division marked in a general way for those screws, circlips, etc. it will receive: for VCRs, top cover, headshield, FL cage, top PCB, bottom PCB, motor brackets, etc. Similarly for TV sets and camcorders: back cover, PCB fixings, control panel; top case, bottom case, deck, screening, etc. If the set is 'shelved' while waiting for spares, still dismantled, the tray compartments can be scooped out into similarly-marked small 'seal-easy' polythene bags for storage with the equipment.

For supporting a VCR while retaining access to underside test-points, or a TV while undergoing tube replacement, it's handy to have carpet-

covered wooden blocks, sized slightly smaller than standard house-bricks, as shown in the photo of Figure 21.7.

Figure 21.7 *Soft-surface blocks find many uses in the workshop*

Hand-tools and service aids

Unless the technician deals with aerials, satellite dishes or washing machines, large and heavy tools are not required for service these days. A basic tool kit may consist of the following, listed roughly in order of importance: an electric soldering iron appropriate to the job in hand; two sizes of crosspoint screwdrivers plus a kit of miniature ('watchmakers') types; ditto, straight-blade; two sizes of side-cutters; medium-size conventional pliers; two sizes of snipe-nose pliers; top-cutters; tweezers; metric and imperial Allen key sets; trimming-tool kit; wire-strippers; miniature nut-spinners, Metric and BA; torx (6-point star) screwdriver set; junior hacksaw; set of miniature files. This list is not exhaustive, but covers most service jobs and situations, and can be expanded as experience dictates; the special tools and jigs mentioned in Chapter 12 will also be needed when dealing with VCR deck mechanics.

Essential service aids are as follows: solder; switch-cleaner and freezer in aerosol dispensers; desoldering braid or pump; graphited and other light greases; silicon grease; light oil; heat-sink compound; PVC tape; non-acid flux; super-glue; epoxy and conventional adhesives; and head-cleaning solvent. Apart from these 'consumable' aids, the following are handy to have: IC test clips, dental mirror, semiconductor equivalents/characteristics books and fixed co-ax attenuators.

Field service toolkit

For the itinerant technician the ideal toolkit is light and portable. Carry a minimum of well-chosen tools and service aids, and avoid steel toolcases, which can cause damage to decor and furniture. It's often useful to carry basic test equipment and documents (job cards, service manuals etc.) in the portable toolkit, and – depending on the type of equipment serviced – some first-line replacement components like fuses, 'top ten' transistors, ICs, remote-control batteries and cassette lamps. Figure 21.8 illustrates a useful form of outside kit.

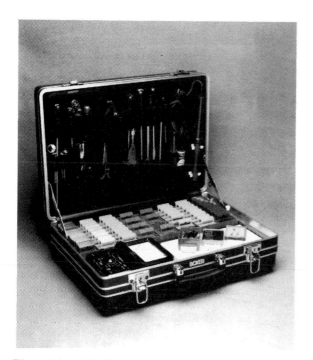

Figure 21.8 *Toolbox and component store combined in a smart case for field service*

Appendix 1

SAFETY, BEAB AND BS415

Electricity, heat and tube implosion are potentially lethal forces, and this should be borne in mind at all times by service technicians. Safety precautions fall into two main categories: protection of the technician while dealing with the equipment; and protection of the user and his property while the equipment is in everyday use.

In the workshop, fit each servicing position with a fused isolating transformer rated at 250W minimum – 500W is better. It is also recommended that the workshop is 'earth-free' in the area of service benches, with grounded objects like radiators, pipes, conduit, trunking etc. either screened by insulating material or well out of reach of the 'operating table'. These precautions only ensure against 'mains-to-earth' shock hazard, so wherever possible keep one hand in your pocket when dealing with high-voltage sections. See also the section of Chapter 10 titled 'handling picture tubes'.

In field servicing, the personal safety risk to the technician is greater. For TV work the best insurance is a portable isolating transformer rated at 200 or 250W; if this cannot be managed, use a roll-up rubber sheet to work on, and keep all test equipment on it. This test equipment (oscilloscope, soldering iron, etc.) should *not* be earthed – check its insulation resistance at intervals. Bear in mind, too, that aerial downleads and plugs are very often effectively grounded, as are the more obvious things like radiators, and these have a dangerously high potential difference to the 'live' section of the equipment, which in some TV designs (especially older ones) includes the 'chassis' and/or common 'ground' line.

The main aspect of safety, however, is the necessity for the equipment to conform to BS415, a copy of the latest form of which can be obtained from BSI Sales, Linford Wood, Milton Keynes MK14 6LE. The onus for complying with BS415, and thus obtaining BEAB approval for the product, rests with the manufacturer, leaving the engineer only to ensure that his actions do not negate the safety features which are built into the design. In the main, this means that all key safety components (marked by the symbol $\langle S \rangle$ or $\triangle$ or shading on the circuit diagram and parts list) should be replaced by types supplied or approved by the manufacturer, and mounted in exactly the same way. All cable ties, clips and cleats removed for access or repair must be replaced, ensuring that there is an air-gap of at least 6mm between live and isolated conductors. All cabinet and cover components must be whole and undamaged, and particular attention should be given to isolating aerial sockets; spacers and insulators associated with user controls and connections; and fixing/assembly screws and bolts.

With a test voltage of 1.5-2kV applied for 1 minute between the mains input leads and user-accessible parts there should be no flashover or sparking; and no more than 2mA d.c. or 700μA a.c. should be able to flow to ground from any user-accessible point in normal operation regardless of the phasing of the a.c. supply connections. Neglect of these check-points lays the repairer open to prosecution for negligence, and liability to damages in the event of death, injury, fire or damage resulting.

Appendix 2

DEALING WITH MOS SEMICONDUCTORS

Many of the ICs used in modern consumer equipment are vulnerable to damage from static charged during handling, installation and storage. Avoid trouble by taking the following precautions:

1 Keep the device in its conductive packing until the last moment before fitting.
2 Store and transport the device in conductive foam or packing.
3 Ground the soldering iron, and the PCB ground land itself before soldering the device into circuit.
4 While handling the device wear a conductive wrist strap connected to ground via a $1M\Omega$ resistor.
5 Disconnect the equipment altogether from the mains while replacing MOS components.
6 If possible keep the device's leadouts electrically strapped together until the soldering operation is complete; this is particularly relevant to MOSFET transistors.

Appendix 3

ADDRESSES

Most of the addresses and phone numbers given below for setmakers are technical departments rather than spares depots. In general they can only deal, for technical information and spares supplies, with their appointed agents and account-holders.

The component suppliers listed here are a representative selection of the larger wholesalers. An increasing trend is for these third-party suppliers to stock a wide range of individual and specific parts for TV and video equipment, working on the setmaker's part numbers – indeed some manufacturers and importers rely entirely on this system for spares supply. Also available from the wholesalers listed here is a wide range of tools, test equipment and service aids.

Manufacturers:

AKAI UK Ltd, Haslemere Heathrow Estate, 12 Silver Jubilee Way, Parkway, Hounslow, Middx TW4 6NF. 01 897 6388

ALBA Electronics, Harvard House, Thames Road, Barking Essex IG1 0HX. 01 594 5533

AMSTRAD Consumer Electronics, Brentwood House, 169 Kings Road, Brentwood Essex. 0277 228888

BANG AND OLUFSEN UK Ltd, Eastbrook Road, Gloucester GL4 7DE. 0452 307377

CANON UK Ltd, Unit 4, Brent Trading Centre, North Circular Road, London NW10. 01 459 1266

DECCA. See *Tatung*.

FERGUSON Ltd, Service Dept, PO Box 1594, Crown Road, Enfield, Middx EN1 1DY. 01 804 7979

FINLUX Finlux House, Braithwaite Street, Leeds, Yorks LS11 9XE. 0532 445645

FISHER UK Ltd, Fisher House, Otterspool Way, Watford, Herts WD2 8JF. 0923 31974

GRUNDIG Ltd, Mill Road, Rugby, Warks CV21 1PR. 0788 77155

HITACHI UK Ltd, Hitachi House, Station Road, Hayes, Middx UB3 4DR. 01 848 8787

ITT UK Ltd, Paycocke Road, Basildon, Essex SS14 3DR. 0268 27788

JVC UK Ltd, 6–8 Priestley Way, Eldonwall Trading Estate, Staples Corner, London NW2 7AF. 01 452 3282

LUXOR. See *SALORA*.

MITSUBISHI UK Ltd, Travellers Lane, Hatfield, Herts AL10 8XB. 07072 76100

NORDMENDE. See *Ferguson*.

PANASONIC UK Ltd, Panasonic House, Willoughby Road, Bracknell, Berks RG12 4FP. 0344 862444

PHILIPS Service, 604 Purley Way, Waddon, Croydon, CR9 4DR. 01 686 0505

PIONEER GB Ltd, 1–6 Fieldway, Greenford, Middx UB6 8UN. 01 575 5757

PYE. See Philips.

SALORA UK Ltd, Bridgemead Close, Westmead Industrial Estate, Westmead, Swindon, Wilts SN5 7YG. 0793 644223

SAMSUNG UK Ltd, Unit 1, Hook Rise Business Centre, 225 Hook Rise South, Surbiton, Surrey KT6 7LD. 01 391 0168

SANYO UK Ltd, Sanyo House, Otterspool Way, Watford, Herts WD2 8JX. 0923 246363

SHARP UK Ltd, Thorp Road, Newton Heath, Manchester M10 9BE. 061 205 2333

SONY UK Ltd, Pipers Way, Thatcham, Newbury, Berks RG13 4LZ. 0635 69500

TATUNG UK Ltd, Stafford Park 10, Telford, Salop TF3 3AB. 0952 290111

TELEFUNKEN Paul Spring Electronics, 6 Oasthouse Way, St. Mary Cray, Orpington, Kent BR5 3PT. 0689 31341

TOSHIBA UK Ltd, Units 6–7, Admiralty Way, Southern Trading Centre, Camberley, Blackwater, Surrey GU15 3DT. 0276 36111

Component Wholesalers

CPC Ltd, 194–200 North Road, Preston, Lancs PR1 1YP. 0772 555034

HRS ELECTRONICS PLC, Garretts Green Lane, Garretts Green, Birmingham, B33 0UE. 021 789 7575

MASTERCARE Components Division, Maylands Court, Maylands Avenue, Hemel Hempstead, Herts HP2 7DE. 0442 232000

MCLELLAND-datatext Ltd, Unit 3A, Bankfield Industrial Estate, Kitson Road, Leeds, Yorks LS10 1NT. 0532 455169

P.V. TUBES, 104 Abbey Street, Accrington, Lancs BB5 1EE. 0254 36521

SEME Ltd, Units 2E/F, Saxby Road Industrial Estate, Melton Mowbray, Leics LE13 1BS. 0664 65392

WILLOW VALE Electronics Ltd, 11 Arkwright Road, Reading, Berks. RG2 0LU. 0734 876444

WIZARD Distributors, Empress Street Works, Empress Street, Manchester, M16 9EN. 0618 725438 or 480060

Service Manual Suppliers

MAURITRON Electronics, 8 Cherry Tree Road, Chinnor, Oxon, OX9 4QY. 0844 51694

TECHNICAL INFO SERVICES, 76 Church Street, Larkhall, Lanarkshire ML9 1HE. 0698 884585

Broadcasters

BBC, Broadcasting House, Portland Place, London W1A 1AA. 01 580 4468 Ext. 2921

IBA, Crawley Court, Winchester, Hants S21 2QA 0962 822444

RTE, Donnybrook, Dublin 4, Eire. Dublin 693111

SYMPTOM INDEX

Note: In general, page numbers below 105 refer to TV faults and numbers above 105 VCR and camera faults. 'Allied' problems are also covered in the text, see individual chapters and sub-titles.